SIX
NOT-SO-EASY
PIECES

SIX
NOT-SO-EASY
PIECES

Einstein's
Relativity, Symmetry,
and Space-Time

RICHARD P.
FEYNMAN

Originally prepared for publication by
Robert B. Leighton and Matthew Sands

New Introduction by Roger Penrose

HELIX BOOKS

Addison-Wesley
Reading, Massachusetts

Many of the designations used by manufacturers and sellers to distinguish their products are claimed as trademarks. Where those designations appear in this book and Addison-Wesley was aware of a trademark claim, the designations have been printed in initial capital letters.

Library of Congress Cataloging-in-Publication Data
Feynman, Richard Phillips.
    Six not-so-easy pieces : Einstein's relativity, symmetry, and
space-time / Richard P. Feynman ; originally prepared for
publication by Robert B. Leighton and Matthew Sands ; new
introduction by Roger Penrose.
      p.   cm.
    Includes index.
    ISBN 0–201–15025–5
    ISBN 0–201–32842–9 (pbk.)
    1. Symmetry (Physics)   2. Special relativity (Physics)   3. Space
and time.   I. Title.
QC793.3.S9F49   1997
530.11—dc21   96-47811
                                               CIP

All text and cover photographs courtesy of the Archives, California Institute of Technology except the photograph on page 92, which is courtesy of the Archives, California Institute of Technology/Yorkshire Television.

Addison-Wesley is an imprint of Addison Wesley Longman, Inc.

Text design by Diane Levy
Set in 11-point Simoncini Garamond by Pagesetters, Inc.

1 2 3 4 5 6 7 8 9 10-DOH-0201009998
First printing, March 1997

First paperback printing, January 1998

Find Helix Books on the World Wide Web site at
http://www.aw.com/gb/

Contents

Contents

Publisher's Note

The unqualified success and popularity of *SIX EASY PIECES* (Addison-Wesley, 1995) sparked a clamor, from the general public, students, and professional scientists alike, for *more* Feynman in book and audio. So we went back to the original *Lectures on Physics* and to the Archives at Caltech to see if there were more "easy" pieces. There were not. But there were many *not-so-easy* lectures that, although they contain some mathematics, are not too difficult for beginning science students; and for the student *and* the lay person, these six lectures are every bit as thrilling, as absorbing, and as much fun as the first six.

Another difference between these not-so-easy pieces and the first six, is that the topics of the first six spanned several fields of physics, from mechanics to thermodynamics to atomic physics. These new six pieces you hold in your hand, however, are focused around a subject which has evoked many of the most revolutionary discoveries and amazing theories of modern physics, from black holes to worm holes, from atomic energy to time warps; we are talking, of course, about Relativity. But even the great Einstein himself, the father of Relativity, could not *explain* the wonders, workings, and fundamental concepts of his own theory as well as could that guy from Noo Yawk, Richard P. Feynman, as reading the chapters or listening to the CDs will prove to you.

Addison Wesley Longman wishes to thank Roger Penrose for

Publisher's Note

his penetrating Introduction to this collection; Brian Hatfield and David Pines for their invaluable advice in the selection of the six lectures; and the California Institute of Technology's Physics Department and Institute Archives, in particular Judith Goodstein, for helping to make this book/CD project happen.

Introduction

To understand why Richard Feynman was such a great teacher, it is important to appreciate his remarkable stature as a scientist. He was indeed one of the outstanding figures of twentieth-century theoretical physics. His contributions to that subject are central to the whole development of the particular way in which quantum theory is used in current cutting-edge research and thus to our present-day pictures of the world. The Feynman path integrals, Feynman diagrams, and Feynman rules are among the very basic tools of the modern theoretical physicist—tools that are necessary for the application of the rules of quantum theory to physical fields (e.g., the quantum theory of electrons, protons, and photons), and which form an essential part of the procedures whereby one makes these rules consistent with the requirements of Einstein's Special Relativity theory. Although none of these ideas is easy to appreciate, Feynman's particular approach always had a deep clarity about it, sweeping away unnecessary complications in what had gone before. There was a close link between his special ability to make progress in research and his particular qualities as a teacher. He had a unique talent that enabled him to cut through the complications that often obscure the essentials of a physical issue and to see clearly into the deep underlying physical principles.

Yet, in the popular conception of Feynman, he is known more for his antics and buffoonery, for his practical jokes, his irreverence towards authority, his bongo-drum performing, his relationships with women, both deep and shallow, his attendance at strip clubs, his attempts, late in life, to reach the obscure country of Tuva in central Asia, and many other schemes. Undoubtedly, he must have been extraordinarily clever, as his lightning quickness at calculation, his exploits involving safe-cracking, outwitting security services, deciphering ancient Mayan texts—not to mention his eventual Nobel Prize—clearly demonstrate. Yet none of this quite conveys the status that he unquestionably has amongst physicists and other scientists, as one of the deepest and most original thinkers of this century.

The distinguished physicist and writer Freeman Dyson, an early collaborator of Feynman's at a time when he was developing his most important ideas, wrote in a letter to his parents in England in the spring of 1948, when Dyson was a graduate student at Cornell University, "Feynman is the young American professor, half genius and half buffoon, who keeps all physicists and their children amused with his effervescent vitality. He has, however, as I have recently learned, a great deal more to him than that. . . ." Much later, in 1988, he would write: "A truer description would have said that Feynman was all genius and all buffoon. The deep thinking and the joyful clowning were not separate parts of a split personality. . . . He was thinking and clowning simultaneously."* Indeed, in his lectures, his wit was spontaneous, and often outrageous. Through it he held his audiences' attention, but never in a way that would distract from the purpose of the lecture, which was the conveying of genuine and deep physical understanding. Through laughter, his audiences could relax and be at ease, rather than feel daunted by what might otherwise be somewhat intimidating mathematical expressions and physical concepts that are tantalizingly difficult to grasp. Yet, although he enjoyed being center stage and was undoubtedly a showman, this was not the purpose of his exposi-

---

* The Dyson quotations are to be found in his book *From Eros to Gaia* (Pantheon Books, New York, 1992) pages 325 and 314, respectively.

Introduction

tions. That purpose was to convey some basic understanding of underlying physical ideas and of the essential mathematical tools that are needed in order to express these ideas properly.

Whereas laughter played a key part of his success in holding an audience's attention, more important to the conveying of understanding was the immediacy of his approach. Indeed, he had an extraordinarily direct no-nonsense style. He scorned airy-fairy philosophizing where it had little physical content. Even his attitude to mathematics was somewhat similar. He had little use for pedantic mathematical niceties, but he had a distinctive mastery of the mathematics that he needed, and could present it in a powerfully transparent way. He was beholden to no one, and would never take on trust what others might maintain to be true without himself coming to an independent judgment. Accordingly, his approach was often strikingly original whether in his research or teaching. And when Feynman's way differed significantly from what had gone before, it would be a reasonably sure bet that Feynman's approach would be the more fruitful one to follow.

Feynman's preferred method of communication was verbal. He did not easily, nor often, commit himself to the printed word. In his scientific papers, the special "Feynman" qualities would certainly come through, though in a somewhat muted form. It was in his lectures that his talents were given full reign. His exceedingly popular "Feynman Lectures" were basically edited transcripts (by Robert B. Leighton and Matthew Sands) of lectures that Feynman gave, and the compelling nature of the text is evident to anyone who reads it. The SIX NOT-SO-EASY PIECES that are presented here are taken from those accounts. Yet, even here, the printed words alone leave something significantly missing. To sense the full excitement that Feynman's lectures exude, I believe that it is important to hear his actual voice. The directness of Feynman's approach, the irreverence, and the humor then become things that we can immediately share in. Fortunately, there are recordings of all the lectures presented in this book, which give us this opportunity—and I strongly recommend, if the opportunity is there, that at least some

Introduction

of these audio versions are listened to first. Once we have heard Feynman's forceful, enthralling, and witty commentary, in the tones of this streetwise New Yorker, we do not forget how he sounds, and it gives us an image to latch on to when we read his words. But whether we actually read the chapters or not, we can share something of the evident thrill that he himself feels as he explores—and continually re-explores—the extraordinary laws that govern the workings of our universe.

The present series of six lectures was carefully chosen to be of a level a little above the six that formed the earlier set of Feynman lectures entitled *Six Easy Pieces* (published by Addison Wesley Longman in 1995). Moreover, they go well together and constitute a superb and compelling account of one of the most important general areas of modern theoretical physics.

This area is *relativity,* which first burst forth into human awareness in the early years of this century. The name of Einstein figures preeminently in the public conception of this field. It was, indeed, Albert Einstein who, in 1905, first clearly enunciated the profound principles which underlie this new realm of physical endeavor. But there were others before him, most notably Hendrik Antoon Lorentz and Henri Poincaré, who had already appreciated most of the basics of the (then) new physics. Moreover, the great scientists Galileo Galilei and Isaac Newton, centuries before Einstein, had already pointed out that in the dynamical theories that they themselves were developing, the physics as perceived by an observer in uniform motion would be identical with that perceived by an observer at rest. The key problem with this had arisen only later, with James Clerk Maxwell's discovery, as published in 1865, of the equations that govern the electric and magnetic fields, and which also control the propagation of light. The implication seemed to be that the relativity principle of Galileo and Newton could no longer hold true; for the speed of light must, by Maxwell's equations, have a definite speed of propagation. Accordingly, an observer at rest is distinguished from those in motion by the fact that only to an observer at rest does the light speed appear to be the same in all

Introduction

directions. The relativity principle of Lorentz, Poincaré, and Einstein differs from that of Galileo and Newton, but it has this same implication: the physics as perceived by an observer in uniform motion is indeed identical with that perceived by an observer at rest.

Yet, in the new relativity, Maxwell's equations *are* consistent with this principle, and the speed of light is measured to have a definite fixed value in every direction, no matter in what direction or with what speed the observer might be moving. How is this magic achieved so that these apparently hopelessly incompatible requirements are reconciled? I shall leave it to Feynman to explain—in his own inimitable fashion.

Relativity is perhaps the first place where the physical power of the mathematical idea of *symmetry* begins to be felt. Symmetry is a familiar idea, but it is less familiar to people how such an idea can be applied in accordance with a set of mathematical expressions. But it is just such a thing that is needed in order to implement the principles of special relativity in a system of equations. In order to be consistent with the relativity principle, whereby physics "looks the same" to an observer in uniform motion as to an observer at rest, there must be a "symmetry transformation" which translates one observer's measured quantities into those of the other. It is a symmetry because the physical laws appear the same to each observer, and "symmetry" after all, asserts that something has the same appearance from two distinct points of view. Feynman's approach to abstract matters of this nature is very down to earth, and he is able to convey the ideas in a way that is accessible to people with no particular mathematical experience or aptitude for abstract thinking.

Whereas relativity pointed the way to additional symmetries that had not been perceived before, some of the more modern developments in physics have shown that certain symmetries, previously thought to be universal, are in fact subtly violated. It came as one of the most profound shocks to the physical community in 1957, as the work of Lee, Yang, and Wu showed, that in certain basic physical processes, the laws satisfied by a physical system are not the same as those satisfied by the mirror reflection of that system. In fact, Feynman had a hand in the development of the physical theory which is able to accommodate this asymmetry. His account here is, accordingly, a dramatic one, as deeper and deeper mysteries of nature gradually unfold.

As physics develops, there are mathematical formalisms that develop with it, and which are needed in order to express the new physical laws. When the mathematical tools are skillfully tuned to their appropriate tasks, they can make the physics seem much simpler than otherwise. The ideas of vector calculus are a case in point. The vector calculus of three dimensions was originally developed to handle the physics of ordinary space, and it provides an invaluable piece of machinery for the expression of physical laws, such as those of Newton, where there is no physically preferred direction in space. To put this another way, the physical laws have a *symmetry* under ordinary rotations in space. Feynman brings home the power of the vector notation and the underlying ideas for expressing such laws.

Relativity theory, however, tells us that *time* should also be brought under the compass of these symmetry transformations, so a *four*-dimensional vector calculus is needed. This calculus is also introduced to us here by Feynman, as it provides the way of understanding how not only time and space must be considered as different aspects of the same four-dimensional structure, but the same is true of energy and momentum in the relativistic scheme.

The idea that the history of the universe should be viewed, physically, as a *four*-dimensional space-time, rather than as a three-dimensional space evolving with time is indeed fundamental to modern physics. It is an idea whose significance is not easy to grasp. Indeed, Einstein himself was not sympathetic to this idea when he first encountered it. The idea of space-time was not, in fact, Einstein's, although, in the popular imagination it is frequently attributed to him. It was the Russian/German geometer Hermann Minkowski, who had been a teacher of Einstein's at the Zurich

Introduction

Polytechnic, who first put forward the idea of four-dimensional space-time in 1908, a few years after Poincaré and Einstein had formulated special relativity theory. In a famous lecture, Minkowski asserted: "Henceforth space by itself, and time by itself, are doomed to fade away into mere shadows, and only a kind of unity between the two will preserve an independent reality."*

Feynman's most influential scientific discoveries, the ones that I have referred to above, stemmed from his own *space-time* approach to quantum mechanics. There is thus no question about the importance of space-time to Feynman's work and to modern physics generally. It is not surprising, therefore, that Feynman is forceful in his promotion of space-time ideas, stressing their physical significance. Relativity is not airy-fairy philosophy, nor is space-time mere mathematical formalism. It is a foundational ingredient of the very universe in which we live.

When Einstein became accustomed to the idea of space-time, he took it completely into his way of thinking. It became an essential part of his extension of special relativity—the relativity theory I have been referring to above that Lorentz, Poincaré, and Einstein introduced—to what is known as *general* relativity. In Einstein's general relativity, the space-time becomes *curved,* and it is able to incorporate the phenomenon of *gravity* into this curvature. Clearly, this is a difficult idea to grasp, and in Feynman's final lecture in this collection, he makes no attempt to describe the full mathematical machinery that is needed for the complete formulation of Einstein's theory. Yet he gives a powerfully dramatic description, with insightful use of intriguing analogies, in order to get the essential ideas across.

In all his lectures, Feynman made particular efforts to preserve accuracy in his descriptions, almost always qualifying what he says when there was any danger that his simplifications or analogies

---

* The Minkowski quote is from the Dover reprint of seminal publications on relativity *The Principle of Relativity* by Einstein, Lorentz, Weyl, and Minkowski (originally Methuen and Co., 1923).

might be misleading or lead to erroneous conclusions. I felt, however, that his simplified account of the Einstein field equation of general relativity did need a qualification that he did not quite give. For in Einstein's theory, the "active" mass which is the source of gravity is not simply the same as the energy (according to Einstein's $E=mc^2$); instead, this source is the energy density *plus the sum of the pressures,* and it is this that is the source of gravity's inward accelerations. With this additional qualification, Feynman's account is superb, and provides an excellent introduction to this most beautiful and self-contained of physical theories.

While Feynman's lectures are unashamedly aimed at those who have aspirations to become physicists—whether professionally or in spirit only—they are undoubtedly accessible also to those with no such aspirations. Feynman strongly believed (and I agree with him) in the importance of conveying an understanding of our universe—according to the perceived basic principles of modern physics—far more widely than can be achieved merely by the teaching provided in physics courses. Even late in his life, when taking part in the investigations of the *Challenger* disaster, he took great pains to show, on national television, that the source of the disaster was something that could be appreciated at an ordinary level, and he performed a simple but convincing experiment on camera showing the brittleness of the shuttle's O-rings in cold conditions.

He was a showman, certainly, sometimes even a clown; but his overriding purpose was always serious. And what more serious purpose can there be than the understanding of the nature of our universe at its deepest levels? At conveying this understanding, Richard Feynman was supreme.

December 1996   ROGER PENROSE

Special Preface
(from *Lectures on Physics*)

Toward the end of his life, Richard Feynman's fame had transcended the confines of the scientific community. His exploits as a member of the commission investigating the space shuttle Challenger disaster gave him widespread exposure; similarly, a best-selling book about his picaresque adventures made him a folk hero almost of the proportions of Albert Einstein. But back in 1961, even before his Nobel Prize increased his visibility to the general public, Feynman was more than merely famous among members of the scientific community—he was legendary. Undoubtedly, the extraordinary power of his teaching helped spread and enrich the legend of Richard Feynman.

He was a truly great teacher, perhaps the greatest of his era and ours. For Feynman, the lecture hall was a theater, and the lecturer a performer, responsible for providing drama and fireworks as well as facts and figures. He would prowl about the front of a classroom, arms waving, "the impossible combination of theoretical physicist and circus barker, all body motion and sound effects," wrote *The New York Times*. Whether he addressed an audience of students, colleagues, or the general public, for those lucky enough to see Feynman lecture in person, the experience was usually unconventional and always unforgettable, like the man himself.

He was the master of high drama, adept at riveting the attention of every lecture-hall audience. Many years ago, he taught a course in Advanced Quantum Mechanics, a large class comprised of a few registered graduate students and most of the Caltech physics faculty. During one lecture, Feynman started explaining how to represent certain complicated integrals diagrammatically: time on this axis, space on that axis, wiggly line for this straight line, etc. Having described what is known to the world of physics as a Feynman diagram, he turned around to face the class, grinning wickedly. "And this is called *THE* diagram!" Feynman had reached the denouement, and the lecture hall erupted with spontaneous applause.

For many years after the lectures that make up this book were given, Feynman was an occasional guest lecturer for Caltech's freshman physics course. Naturally, his appearances had to be kept secret so there would be room left in the hall for the registered students. At one such lecture the subject was curved-space time, and Feynman was characteristically brilliant. But the unforgettable moment came at the beginning of the lecture. The supernova of 1987 has just been discovered, and Feynman was very excited about it. He said, "Tycho Brahe had his supernova, and Kepler had his. Then there weren't any for 400 years. But now I have mine." The class fell silent, and Feynman continued on. "There are 10^{11} stars in the galaxy. That used to be a *huge* number. But it's only a hundred billion. It's less than the national deficit! We used to call them astronomical numbers. Now we should call them economical numbers." The class dissolved in laughter, and Feynman, having captured his audience, went on with his lecture.

Showmanship aside, Feynman's pedagogical technique was simple. A summation of his teaching philosophy was found among his papers in the Caltech archives, in a note he had scribbled to himself while in Brazil in 1952:

> "First figure out why you want the students to learn the subject and what you want them to know, and the method will result more or less by common sense."

What came to Feynman by "common sense" were often brilliant twists that perfectly captured the essence of his point. Once, during a public lecture, he was trying to explain why one must not verify an idea using the same data that suggested the idea in the first place. Seeming to wander off the subject, Feynman began talking about license plates. "You know, the most amazing thing happened to me tonight. I was coming here, on the way to the lecture, and I came in through the parking lot. And you won't believe what happened. I saw a car with the license plate ARW 357. Can you imagine? Of all the millions of license plates in the state, what was the chance that I would see that particular one tonight? Amazing!" A point that even many scientists fail to grasp was made clear through Feynman's remarkable "common sense."

In 35 years at Caltech (from 1952 to 1987), Feynman was listed as teacher of record for 34 courses. Twenty-five of them were advanced graduate courses, strictly limited to graduate students, unless undergraduates asked permission to take them (they often did, and permission was nearly always granted). The rest were mainly introductory graduate courses. Only once did Feynman teach courses purely for undergraduates, and that was the celebrated occasion in the academic years 1961 to 1962 and 1962 to 1963, with a brief reprise in 1964, when he gave the lectures that were to become *The Feynman Lectures on Physics*.

At the time there was a consensus at Caltech that freshman and sophomore students were getting turned off rather than spurred on by their two years of compulsory physics. To remedy the situation, Feynman was asked to design a series of lectures to be given to the students over the course of two years, first to freshmen, and then to the same class as sophomores. When he agreed, it was immediately decided that the lectures should be transcribed for publication. That job turned out to be far more difficult than anyone had imagined. Turning out publishable books required a tremendous amount of work on the part of his colleagues, as well as Feynman himself, who did the final editing of every chapter.

And the nuts and bolts of running a course had to be addressed. This task was greatly complicated by the fact that Feynman had only a vague outline of what he wanted to cover. This meant that no one knew what Feynman would say until he stood in front of a lecture hall filled with students and said it. The Caltech professors who assisted him would then scramble as best they could to handle mundane details, such as making up homework problems.

Why did Feynman devote more than two years to revolutionizing the way beginning physics was taught? One can only speculate, but there were probably three basic reasons. One is that he loved to have an audience, and this gave him a bigger theater than he usually had in graduate courses. The second was that he genuinely cared about students, and he simply thought that teaching freshmen was an important thing to do. The third and perhaps most important reason was the sheer challenge of reformulating physics, as he understood it, so that it could be presented to young students. This was his specialty, and was the standard by which he measured whether something was really understood. Feynman was once asked by a Caltech faculty member to explain why spin 1/2 particles obey Fermi-Dirac statistics. He gauged his audience perfectly and said, "I'll prepare a freshman lecture on it." But a few days later he returned and said, "You know, I couldn't do it. I couldn't reduce it to the freshman level. That means we really don't understand it."

This specialty of reducing deep ideas to simple, understandable terms is evident throughout *The Feynman Lectures on Physics*, but nowhere more so than in his treatment of quantum mechanics. To aficionados, what he has done is clear. He has presented, to beginning students, the path integral method, the technique of his own devising that allowed him to solve some of the most profound problems in physics. His own work using path integrals, among other achievements, led to the 1965 Nobel Prize that he shared with Julian Schwinger and Sin-Itero Tomanaga.

Through the distant veil of memory, many of the students and faculty attending the lectures have said that having two years of physics with Feynman was the experience of a lifetime. But that's not how it seemed at the time. Many of the students dreaded the class, and as the course wore on, attendance by the registered students started dropping alarmingly. But at the same time, more and more faculty and graduate students started attending. The room stayed full, and Feynman may never have known he was losing some of his intended audience. But even in Feynman's view, his pedagogical endeavor did not succeed. He wrote in the 1963 preface to the *Lectures*: "I don't think I did very well by the students." Rereading the books, one sometimes seems to catch Feynman looking over his shoulder, not at his young audience, but directly at his colleagues, saying, "Look at that! Look how I finessed that point! Wasn't that clever?" But even when he thought he was explaining things lucidly to freshmen or sophomores, it was not really they who were able to benefit most from what he was doing. It was his peers—scientists, physicists, and professors—who would be the main beneficiaries of his magnificent achievement, which was nothing less than to see physics through the fresh and dynamic perspective of Richard Feynman.

Feynman was more than a great teacher. His gift was that he was an extraordinary teacher of teachers. If the purpose in giving *The Feynman Lectures on Physics* was to prepare a roomful of undergraduate students to solve examination problems in physics, he cannot be said to have succeeded particularly well. Moreover, if the intent was for the books to serve as introductory college textbooks, he cannot be said to have achieved his goal. Nevertheless, the books have been translated into ten foreign languages and are available in four bilingual editions. Feynman himself believed that his most important contribution to physics would not be QED, or the theory of superfluid helium, or polarons, or partons. His foremost contribution would be the three red books of *The Feynman Lectures on Physics*. That belief justifies this commemorative issue of these celebrated books.

DAVID L. GOODSTEIN
GERRY NEUGEBAUER
April 1989　　California Institute of Technology

Feynman's Preface
(from *Lectures on Physics*)

These are the lectures in physics that I gave last year and the year before to the freshman and sophomore classes at Caltech. The lectures are, of course, not verbatim—they have been edited, sometimes extensively and sometimes less so. The lectures form only part of the complete course. The whole group of 180 students gathered in a big lecture room twice a week to hear these lectures and then they broke up into small groups of 15 to 20 students in recitation sections under the guidance of a teaching assistant. In addition, there was a laboratory session once a week.

The special problem we tried to get at with these lectures was to maintain the interest of the very enthusiastic and rather smart students coming out of the high schools and into Caltech. They have heard a lot about how interesting and exciting physics is—the theory of relativity, quantum mechanics, and other modern ideas. By the end of two years of our previous course, many would be very discouraged because there were really very few grand, new, modern ideas presented to them. They were made to study inclined planes, electrostatics, and so forth, and after two years it was quite stultifying. The problem was whether or not we could make a course which would save the more advanced and excited student by maintaining his enthusiasm.

The lectures here are not in any way meant to be a survey course, but are very serious. I thought to address them to the most intelligent in the class and to make sure, if possible, that even the most intelligent student was unable to completely encompass everything that was in the lectures—by putting in suggestions of applications of the ideas and concepts in various directions outside the main line of attack. For this reason, though, I tried very hard to make all the statements as accurate as possible, to point out in every case where the equations and ideas fitted into the body of physics, and how— when they learned more—things would be modified. I also felt that for such students it is important to indicate what it is that they should—if they are sufficiently clever—be able to understand by deduction from what has been said before, and what is being put in as something new. When new ideas came in, I would try either to deduce them if they were deducible, or to explain that it *was* a new idea which hadn't any basis in terms of things they had already learned and what was not supposed to be provable—but was just added in.

At the start of these lectures, I assumed that the students knew something when they came out of high school—such things as geometrical optics, simple chemistry ideas, and so on. I also didn't see that there was any reason to make the lectures in a definite order, in the sense that I would not be allowed to mention something until I was ready to discuss it in detail. There was a great deal of mention of things to come, without complete discussions. These more complete discussions would come later when the preparation became more advanced. Examples are the discussions of inductance, and of energy levels, which are at first brought in in a very qualitative way and are later developed more completely.

At the same time that I was aiming at the more active student, I also wanted to take care of the fellow for whom the extra fireworks and side applications are merely disquieting and who cannot be expected to learn most of the material in the lecture at all. For such a student, I wanted there to be at least a central core or backbone of material which he *could* get. Even if he didn't understand everything in a lecture, I hoped he wouldn't get nervous. I didn't expect him to understand everything, but only the central and most direct features. It takes, of course, a certain intelligence on his part to see

Feynman's Preface

which are the central theorems and central ideas, and which are the more advanced side issues and applications which he may understand only in later years.

In giving these lectures there was one serious difficulty: in the way the course was given, there wasn't any feedback from the students to the lecturer to indicate how well the lectures were going over. This is indeed a very serious difficulty, and I don't know how good the lectures really are. The whole thing was essentially an experiment. And if I did it again I wouldn't do it the same way—I hope I *don't* have to do it again! I think, though, that things worked out—so far as the physics is concerned—quite satisfactorily in the first year.

In the second year I was not so satisfied. In the first part of the course, dealing with electricity and magnetism, I couldn't think of any really unique or different way of doing it—of any way that would be particularly more exciting than the usual way of presenting it. So I don't think I did very much in the lectures on electricity and magnetism. At the end of the second year I had originally intended to go on, after the electricity and magnetism, by giving some more lectures on the properties of materials, but mainly to take up things like fundamental modes, solutions of the diffusion equation, vibrating systems, orthogonal functions, . . . developing the first stages of what are usually called "the mathematical methods of physics." In retrospect, I think that if I were doing it again I would go back to that original idea. But since it was not planned that I would be giving these lectures again, it was suggested that it might be a good idea to try to give an introduction to the quantum mechanics—what you will find in Volume III.

It is perfectly clear that students who will major in physics can wait until their third year for quantum mechanics. On the other hand, the argument was made that many of the students in our course study physics as a background for their primary interest in other fields. And the usual way of dealing with quantum mechanics makes that subject almost unavailable for the great majority of students because they have to take so long to learn it. Yet, in its real applications—especially in its more complex applications, such as in electrical engineering and chemistry—the full machinery of the differential equation approach is not actually used. So I tried to describe the principles of quantum mechanics in a way which wouldn't require that one first know the mathematics of partial differential equations. Even for a physicist I think that is an interesting thing to try to do—to present quantum mechanics in this reverse fashion—for several reasons which may be apparent in the lectures themselves. However, I think that the experiment in the quantum mechanics part was not completely successful—in large part because I really did not have enough time at the end (I should, for instance, have had three or four more lectures in order to deal more completely with such matters as energy bands and the spatial dependence of amplitudes). Also, I had never presented the subject this way before, so the lack of feedback was particularly serious. I now believe the quantum mechanics should be given at a later time. Maybe I'll have a chance to do it again someday. Then I'll do it right.

The reason there are no lectures on how to solve problems is because there were recitation sections. Although I did put in three lectures in the first year on how to solve problems, they are not included here. Also there was a lecture on inertial guidance which certainly belongs after the lecture on rotating systems, but which was, unfortunately, omitted. The fifth and sixth lectures are actually due to Matthew Sands, as I was out of town.

The question, of course, is how well this experiment has succeeded. My own point of view—which, however, does not seem to be shared by most of the people who worked with the students—is pessimistic. I don't think I did very well by the students. When I look at the way the majority of the students handled the problems on the examinations, I think that the system is a failure. Of course, my friends point out to me that there were one or two dozen students who—very surprisingly—understood almost everything in all of the lectures, and who were quite active in working with the material and worrying about the many points in an excited and interested way. These people have now, I believe, a first-rate back-

Feynman's Preface

ground in physics—and they are, after all, the ones I was trying to get at. But then, "The power of instruction is seldom of much efficacy except in those happy dispositions where it is almost superfluous." (Gibbon)

Still, I didn't want to leave any student completely behind, as perhaps I did. I think one way we could help the students more would be by putting more hard work into developing a set of problems which would elucidate some of the ideas in the lectures. Problems give a good opportunity to fill out the material of the lectures and make more realistic, more complete, and more settled in the mind the ideas that have been exposed.

I think, however, that there isn't any solution to this problem of education other than to realize that the best teaching can be done only when there is a direct individual relationship between a student and a good teacher—a situation in which the student discusses the ideas, thinks about the things, and talks about the things. It's impossible to learn very much by simply sitting in a lecture, or even by simply doing problems that are assigned. But in our modern times we have so many students to teach that we have to try to find some substitute for the ideal. Perhaps my lectures can make some contribution. Perhaps in some small place where there are individual teachers and students, they may get some inspiration or some ideas from the lectures. Perhaps they will have fun thinking them through—or going on to develop some of the ideas further.

June 1963   RICHARD P. FEYNMAN

SIX
NOT-SO-EASY
PIECES

One

VECTORS

1-1 Symmetry in physics

◊ In this chapter we introduce a subject that is technically known
  in physics as *symmetry in physical law*. The word "symmetry" is
used here with a special meaning, and therefore needs to be
defined. When is a thing symmetrical—how can we define it?
When we have a picture that is symmetrical, one side is somehow
the same as the other side. Professor Hermann Weyl has given this
definition of symmetry: a thing is symmetrical if one can subject it to
a certain operation and it appears exactly the same after the operation. For instance, if we look at a vase that is left-and-right symmetrical, then turn it 180° around the vertical axis, it looks the same. We
shall adopt the definition of symmetry in Weyl's more general form,
and in that form we shall discuss symmetry of physical laws.

Suppose we build a complex machine in a certain place, with a lot
of complicated interactions, and balls bouncing around with forces
between them, and so on. Now suppose we build exactly the same
kind of equipment at some other place, matching part by part, with
the same dimensions and the same orientation, everything the same
only displaced laterally by some distance. Then, if we start the two
machines in the same initial circumstances, in exact correspondence, we ask: will one machine behave exactly the same as the
other? Will it follow all the motions in exact parallelism? Of course
the answer may well be *no*, because if we choose the wrong place for

1

our machine it might be inside a wall and interferences from the wall would make the machine not work.

All of our ideas in physics require a certain amount of common sense in their application; they are not purely mathematical or abstract ideas. We have to understand what we mean when we say that the phenomena are the same when we move the apparatus to a new position. We mean that we move everything that we believe is relevant; if the phenomenon is not the same, we suggest that something relevant has not been moved, and we proceed to look for it. If we never find it, then we claim that the laws of physics do not have this symmetry. On the other hand, we may find it—we expect to find it—if the laws of physics do have this symmetry; looking around, we may discover, for instance, that the wall is pushing on the apparatus. The basic question is, if we define things well enough, if all the essential forces are included inside the apparatus, if all the relevant parts are moved from one place to another, will the laws be the same? Will the machinery work the same way?

It is clear that what we want to do is to move all the equipment and *essential* influences, but not *everything* in the world—planets, stars, and all—for if we do that, we have the same phenomenon again for the trivial reason that we are right back where we started. No, we cannot move *everything*. But it turns out in practice that with a certain amount of intelligence about what to move, the machinery will work. In other words, if we do not go inside a wall, if we know the origin of the outside forces, and arrange that those are moved too, then the machinery *will* work the same in one location as in another.

1-2 *Translations*

We shall limit our analysis to just mechanics, for which we now have sufficient knowledge. In previous chapters we have seen that the laws of mechanics can be summarized by a set of three equations for each particle:

$$m(d^2x/dt^2) = F_x, \quad m(d^2y/dt^2) = F_y, \quad m(d^2z/dt^2) = F_z. \quad (1.1)$$

Now this means that there exists a way to *measure x, y,* and *z* on three perpendicular axes, and the forces along those directions, such that these laws are true. These must be measured from some origin, but *where do we put the origin?* All that Newton would tell us at first is that there *is* some place that we can measure from, perhaps the center of the universe, such that these laws are correct. But we can show immediately that we can never find the center, because if we use some other origin it would make no difference. In other words, suppose that there are two people—Joe, who has an origin in one place, and Moe, who has a parallel system whose origin is somewhere else (Figure 1-1). Now when Joe measures the location of the point in space, he finds it at *x, y,* and *z* (we shall usually leave *z* out because it is too confusing to draw in a picture). Moe, on the other hand, when measuring the same point, will obtain a different *x* (in order to distinguish it, we will call it *x'*), and in principle a different *y*, although in our example they are numerically equal. So we have

$$x' = x - a, \quad y' = y, \quad z' = z. \tag{1.2}$$

Now in order to complete our analysis we must know what Moe would obtain for the forces. The force is supposed to act along some line, and by the force in the *x*-direction we mean the part of the total which is in the *x*-direction, which is the magnitude of the force times this cosine of its angle with the *x*-axis. Now we see that

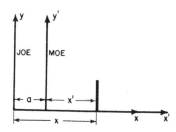

Figure 1-1. Two parallel coordinate systems.

Moe would use exactly the same projection as Joe would use, so we have a set of equations

$$F_{x'} = F_x, \quad F_{y'} = F_y, \quad F_{z'} = F_z. \tag{1.3}$$

These would be the relationships between quantities as seen by Joe and Moe.

The question is, if Joe knows Newton's laws, and if Moe tries to write down Newton's laws, will they also be correct for him? Does it make any difference from which origin we measure the points? In other words, assuming that equations (1.1) are true, and the Eqs. (1.2) and (1.3) give the relationship of the measurements, is it or is it not true that

(a) $m(d^2x'/dt^2) = F_{x'}$,

(b) $m(d^2y'/dt^2) = F_{y'}$, $\qquad\qquad$ (1.4)

(c) $m(d^2z'/dt^2) = F_{z'}$?

In order to test these equations we shall differentiate the formula for x' twice. First of all

$$\frac{dx'}{dt} = \frac{d}{dt}(x - a) = \frac{dx}{dt} - \frac{da}{dt}.$$

Now we shall assume that Moe's origin is fixed (not moving) relative to Joe's; therefore a is a constant and $da/dt = 0$, so we find that

$$dx'/dt = dx/dt$$

and therefore

$$d^2x'/dt^2 = d^2x/dt^2;$$

therefore we know that Eq. (1.4a) becomes

$$m(d^2x/dt^2) = F_{x'}.$$

(We also suppose that the masses measured by Joe and Moe are equal.) Thus the acceleration times the mass is the same as the other fellow's. We have also found the formula for $F_{x'}$, for, substituting from Eq. (1.1), we find that

$$F_{x'} = F_x.$$

Vectors

Therefore the laws as seen by Moe appear the same; he can write Newton's laws too, with different coordinates, and they will still be right. That means that there is no unique way to define the origin of the world, because the laws will appear the same, from whatever position they are observed.

This is also true: if there is a piece of equipment in one place with a certain kind of machinery in it, the same equipment in another place will behave in the same way. Why? Because one machine, when analyzed by Moe, has exactly the same equations as the other one, analyzed by Joe. Since the *equations* are the same, the *phenomena* appear the same. So the proof that an apparatus in a new position behaves the same as it did in the old position is the same as the proof that the equations when displaced in space reproduce themselves. Therefore we say that *the laws of physics are symmetrical for translational displacements*, symmetrical in the sense that the laws do not change when we make a translation of our coordinates. Of course it is quite obvious intuitively that this is true, but it is interesting and entertaining to discuss the mathematics of it.

1-3 Rotations

The above is the first of a series of ever more complicated propositions concerning the symmetry of a physical law. The next proposition is that it should make no difference in which *direction* we choose the axes. In other words, if we build a piece of equipment in some place and watch it operate, and nearby we build the same kind of apparatus but put it up on an angle, will it operate in the same way? Obviously it will not if it is a grandfather clock, for example! If a pendulum clock stands upright, it works fine, but if it is tilted the pendulum falls against the side of the case and nothing happens. The theorem is then false in the case of the pendulum clock, unless we include the earth, which is pulling on the pendulum. Therefore we can make a prediction about pendulum clocks if we believe in the symmetry of physical law for rotation: something else is involved in the operation of a pendulum clock besides the machinery of the clock, something outside it that we should look for. We may also predict that pendulum clocks will not work the same way when located in different places relative to this mysterious source of asymmetry, perhaps the earth. Indeed, we know that a pendulum clock up in an artificial satellite, for example, would not tick either, because there is no effective force, and on Mars it would go at a different rate. Pendulum clocks *do* involve something more than just the machinery inside, they involve something on the outside. Once we recognize this factor, we see that we must turn the earth along with the apparatus. Of course we do not have to worry about that, it is easy to do; one simply waits a moment or two and the earth turns; then the pendulum clock ticks again in the new position the same as it did before. While we are rotating in space our angles are always changing, absolutely; this change does not seem to bother us very much, for in the new position we seem to be in the same condition as in the old. This has a certain tendency to confuse one, because it is true that in the new turned position the laws are the same as in the unturned position, but it is *not* true that *as we turn* a thing it follows the same laws as it does when we are not turning it. If we perform sufficiently delicate experiments, we can tell that the earth *is rotating*, but not that it *had rotated*. In other words, we cannot locate its angular position, but we can tell that it is changing.

Now we may discuss the effects of angular orientation upon physical laws. Let us find out whether the same game with Joe and Moe works again. This time, to avoid needless complication, we shall suppose that Joe and Moe use the same origin (we have already shown that the axes can be moved by translation to another place). Assume that Moe's axes have rotated relative to Joe's by an angle θ. The two coordinate systems are shown in Figure 1-2, which is restricted to two dimensions. Consider any point P having coordinates (x, y) in Joe's system and (x', y') in Moe's system. We shall begin, as in the previous case, by expressing the coordinates x' and y' in terms of x, y, and θ. To do so, we first drop perpendiculars from P to all four axes and draw AB perpendicular to PQ. Inspection of the figure shows that x' can be written as the sum of two

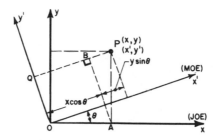

Figure 1-2. Two coordinate systems having different angular orientations.

lengths along the x'-axis, and y' as the difference of two lengths
along AB. All these lengths are expressed in terms of x, y, and θ in
equations (1.5), to which we have added an equation for the third
dimension.

$$x' = x \cos \theta + y \sin \theta,$$
$$y' = y \cos \theta - x \sin \theta, \tag{1.5}$$
$$z' = z.$$

The next step is to analyze the relationship of forces as seen by
the two observers, following the same general method as before. Let
us assume that a force F, which has already been analyzed as having
components F_x and F_y (as seen by Joe), is acting on a particle of
mass m, located at point P in Figure 1-2. For simplicity, let us move
both sets of axes so that the origin is at P, as shown in Figure 1-3.
Moe sees the components of F along his axes as $F_{x'}$ and $F_{y'}$. F_x has
components along both the x'- and y'-axes, and F_y likewise has
components along both these axes. To express $F_{x'}$ in terms of F_x and
F_y, we sum these components along the x'-axis, and in a like manner
we can express $F_{y'}$ in terms of F_x and F_y. The results are

$$F_{x'} = F_x \cos \theta + F_y \sin \theta,$$
$$F_{y'} = F_y \cos \theta - F_x \sin \theta, \tag{1.6}$$
$$F_{z'} = F_z.$$

SIX NOT-SO-EASY PIECES

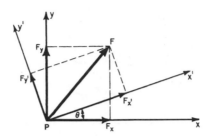

Figure 1-3. Components of a force in the two systems.

It is interesting to note an accident of sorts, which is of extreme importance: the formulas (1.5) and (1.6), for coordinates of P and components of F, respectively, *are of identical form*.

As before, Newton's laws are assumed to be true in Joe's system, and are expressed by equations (1.1). The question, again, is whether Moe can apply Newton's laws—will the results be correct for his system of rotated axes? In other words, if we assume that Eqs. (1.5) and (1.6) give the relationship of the measurements, is it true or not true that

$$m(d^2x'/dt^2) = F_{x'},$$
$$m(d^2y'/dt^2) = F_{y'},$$
$$m(d^2z'/dt^2) = F_{z'}? \qquad (1.7)$$

To test these equations, we calculate the left and right sides independently, and compare the results. To calculate the left sides, we multiply equations (1.5) by m, and differentiate twice with respect to time, assuming the angle θ to be constant. This gives

$$m(d^2x'/dt^2) = m(d^2x/dt^2) \cos\theta + m(d^2y/dt^2) \sin\theta,$$
$$m(d^2y'/dt^2) = m(d^2y/dt^2) \cos\theta - m(d^2x/dt^2) \sin\theta, \qquad (1.8)$$
$$m(d^2z'/dt^2) = m(d^2z/dt^2).$$

We calculate the right sides of equations (1.7) by substituting equations (1.1) into equations (1.6). This gives

Vectors

$$F_{x'} = m(d^2x/dt^2) \cos \theta + m(d^2y/dt^2) \sin \theta,$$
$$F_{y'} = m(d^2y/dt^2) \cos \theta - m(d^2x/dt^2) \sin \theta, \qquad (1.9)$$
$$F_{z'} = m(d^2z/dt^2).$$

Behold! The right sides of Eqs. (1.8) and (1.9) are identical, so we conclude that if Newton's laws are correct on one set of axes, they are also valid on any other set of axes. This result, which has now been established for both translation and rotation of axes, has certain consequences: first, no one can claim his particular axes are unique, but of course they can be more *convenient* for certain particular problems. For example, it is handy to have gravity along one axis, but this is not physically necessary. Second, it means that any piece of equipment which is completely self-contained, with all the force-generating equipment completely inside the apparatus, would work the same when turned at an angle.

1-4 Vectors

Not only Newton's laws, but also the other laws of physics, so far as we know today, have the two properties which we call invariance (or symmetry) under translation of axes and rotation of axes. These properties are so important that a mathematical technique has been developed to take advantage of them in writing and using physical laws.

The foregoing analysis involved considerable tedious mathematical work. To reduce the details to a minimum in the analysis of such questions, a very powerful mathematical machinery has been devised. This system, called *vector analysis*, supplies the title of this chapter; strictly speaking, however, this is a chapter on the symmetry of physical laws. By the methods of the preceding analysis we were able to do everything required for obtaining the results that we sought, but in practice we should like to do things more easily and rapidly, so we employ the vector technique.

We begin by noting some characteristics of two kinds of quantities that are important in physics. (Actually there are more than two, but let us start out with two.) One of them, like the number of potatoes in a sack, we call an ordinary quantity, or an undirected quantity, or a *scalar*. Temperature is an example of such a quantity. Other quantities that are important in physics do have direction, for instance velocity: we have to keep track of which way a body is going, not just its speed. Momentum and force also have direction, as does displacement: when someone steps from one place to another in space, we can keep track of how far he went, but if we wish also to know *where* he went, we have to specify a direction.

All quantities that have a direction, like a step in space, are called *vectors*.

A vector is three numbers. In order to represent a step in space, say from the origin to some particular point P whose location is (x, y, z), we really need three numbers, but we are going to invent a single mathematical symbol, **r**, which is unlike any other mathematical symbols we have so far used.* It is *not* a single number, it represents *three* numbers: x, y, and z. It means three numbers, but not really only *those* three numbers, because if we were to use a different coordinate system, the three numbers would be changed to x', y', and z'. However, we want to keep our mathematics simple and so we are going to use the *same mark* to represent the three numbers (x, y, z) and the three numbers (x', y', z'). That is, we use the same mark to represent the first set of three numbers for one coordinate system, but the second set of three numbers if we are using the other coordinate system. This has the advantage that when we change the coordinate system, we do not have to change the letters of our equations. If we write an equation in terms of x, y, z, and then use another system, we have to change to x', y', z', but we shall just write **r**, with the convention that it represents (x, y, z) if we use one set of axes, or (x', y', z') if we use another set of axes, and so on. The three numbers which describe the quantity in a given coordinate system are called the *components* of the vector in the

---

* In type, vectors are represented by boldface; in handwritten form an arrow is used: $\bar{r}$.

Vectors

direction of the coordinate axes of that system. That is, we use the same symbol for the three letters that correspond to the *same object, as seen from different axes.* The very fact that we can say "the same object" implies a physical intuition about the reality of a step in space, that is independent of the components in terms of which we measure it. So the symbol **r** will represent the same thing no matter how we turn the axes.

Now suppose there is another directed physical quantity, any other quantity, which also has three numbers associated with it, like force, and these three numbers change to three other numbers by a certain mathematical rule, if we change the axes. It must be the same rule that changes (x, y, z) into (x', y', z'). In other words, any physical quantity associated with three numbers which transform as do the components of a step in space is a vector. An equation like

$$\mathbf{F} = \mathbf{r}$$

would thus be true in *any* coordinate system if it were true in one. This equation, of course, stands for the three equations

$$F_x = x, \quad F_y = y, \quad F_z = z,$$

or, alternatively, for

$$F_{x'} = x', \quad F_{y'} = y', \quad F_{z'} = z'.$$

The fact that a physical relationship can be expressed as a vector equation assures us the relationship is unchanged by a mere rotation of the coordinate system. That is the reason why vectors are so useful in physics.

Now let us examine some of the properties of vectors. As examples of vectors we may mention velocity, momentum, force, and acceleration. For many purposes it is convenient to represent a vector quantity by an arrow that indicates the direction in which it is acting. Why can we represent force, say, by an arrow? Because it has the same mathematical transformation properties as a "step in space." We thus represent it in a diagram as if it were a step, using a scale such that one unit of force, or one newton, corresponds to a certain convenient length. Once we have done this, all forces can be represented as lengths, because an equation like

$$\mathbf{F} = k\mathbf{r},$$

where k is some constant, is a perfectly legitimate equation. Thus we can always represent forces by lines, which is very convenient, because once we have drawn the line we no longer need the axes. Of course, we can quickly calculate the three components as they change upon turning the axes, because that is just a geometric problem.

1-5 Vector algebra

Now we must describe the laws, or rules, for combining vectors in various ways. The first such combination is the *addition* of two vectors: suppose that **a** is a vector which in some particular coordinate system has the three components (a_x, a_y, a_z), and that **b** is another vector which has the three components (b_x, b_y, b_z). Now let us invent three new numbers $(a_x + b_x, a_y + b_y, a_z + b_z)$. Do these form a vector? "Well," we might say, "they are three numbers, and every three numbers form a vector." No, *not* every three numbers form a vector! In order for it to be a vector, not only must there be three numbers, but these must be associated with a coordinate system in such a way that if we turn the coordinate system, the three numbers "revolve" on each other, get "mixed up" in each other, by the precise laws we have already described. So the question is, if we now rotate the coordinate system so that (a_x, a_y, a_z) become $(a_{x'}, a_{y'}, a_{z'})$ and (b_x, b_y, b_z) become $(b_{x'}, b_{y'}, b_{z'})$, what do $(a_x + b_x, a_y + b_y, a_z + b_z)$ become? Do they become $(a_{x'} + b_{x'}, a_{y'} + b_{y'}, a_{z'} + b_{z'})$ or not? The answer is, of course, yes, because the prototype transformations of Eq. (1.5) constitute what we call a *linear* transformation. If we apply those transformations to a_x and b_x to get $a_{x'} + b_{x'}$, we find that the transformed $a_x + b_x$ is indeed the same as $a_{x'} + b_{x'}$. When **a** and **b** are "added together" in this sense, they will form a vector which we may call **c**. We would write this as

Vectors

$$c = a + b.$$

Now **c** has the interesting property

$$c = b + a,$$

as we can immediately see from its components. Thus also,

$$a + (b + c) = (a + b) + c.$$

We can add vectors in any order.

What is the geometric significance of **a** + **b**? Suppose that **a** and **b** were represented by lines on a piece of paper, what would **c** look like? This is shown in Figure 1-4. We see that we can add the components of **b** to those of **a** most conveniently if we place the rectangle representing the components of **b** next to that representing the components of **a** in the manner indicated. Since **b** just "fits" into its rectangle, as does **a** into its rectangle, this is the same as putting the "tail" of **b** on the "head" of **a**, the arrow from the "tail" of **a** to the "head" of **b** being the vector **c**. Of course, if we added **a** to **b** the other way around, we would put the "tail" of **a** on the "head" of **b**, and by the geometrical properties of parallelograms we would get the same result for **c**. Note that vectors can be added in this way without reference to any coordinate axes.

Suppose we multiply a vector by a number α, what does this mean? We *define* it to mean a new vector whose components are

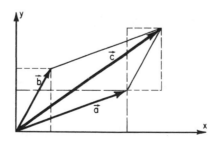

Figure 1-4. The addition of vectors.

αa_x, αa_y, and αa_z. We leave it as a problem for the student to prove that it *is* a vector.

Now let us consider vector subtraction. We may define subtraction in the same way as addition, but instead of adding, we subtract the components. Or we might define subtraction by defining a negative vector, $-\mathbf{b} = -1\mathbf{b}$, and then we would add the components. It comes to the same thing. The result is shown in Figure 1-5. This figure shows $\mathbf{d} = \mathbf{a} - \mathbf{b} = \mathbf{a} + (-\mathbf{b})$; we also note that the difference $\mathbf{a} - \mathbf{b}$ can be found very easily from $\mathbf{a}$ and $\mathbf{b}$ by using the equivalent relation $\mathbf{a} = \mathbf{b} + \mathbf{d}$. Thus the difference is even easier to find than the sum: we just draw the vector from $\mathbf{b}$ to $\mathbf{a}$, to get $\mathbf{a} - \mathbf{b}$!

Next we discuss velocity. Why is velocity a vector? If position is given by the three coordinates (x, y, z), what is the velocity? The velocity is given by dx/dt, dy/dt, and dz/dt. Is that a vector, or not? We can find out by differentiating the expressions in Eq. (1.5) to find out whether dx'/dt *transforms* in the right way. We see that the components dx/dt and dy/dt *do* transform according to the same law as x and y, and therefore the time derivative is a vector. So the velocity *is* a vector. We can write the velocity in an interesting way as

$$\mathbf{v} = d\mathbf{r}/dt.$$

What the velocity is, and why it is a vector, can also be understood more pictorially: How far does a particle move in a short time Δt?

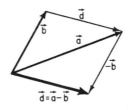

$$\vec{d} = \vec{a} - \vec{b}$$

Figure 1-5. The subtraction of vectors.

Vectors

Answer: $\Delta\mathbf{r}$, so if a particle is "here" at one instant and "there" at another instant, then the vector difference of the positions $\Delta\mathbf{r} = \mathbf{r}_2 - \mathbf{r}_1$, which is in the direction of motion shown in Figure 1-6, divided by the time interval $\Delta t = t_2 - t_1$, is the "average velocity" vector.

In other words, by vector velocity we mean the limit, as Δt goes to 0, of the difference between the radius vectors at the time $t + \Delta t$ and the time t, divided by Δt:

$$\mathbf{v} = \lim_{\Delta t \to 0} (\Delta\mathbf{r}/\Delta t) = d\mathbf{r}/dt. \qquad (1.10)$$

Thus velocity is a vector because it is the difference of two vectors. It is also the right definition of velocity because its components are dx/dt, dy/dt, and dz/dt. In fact, we see from this argument that if we differentiate *any* vector with respect to time we produce a new vector. So we have several ways of producing new vectors: (1) multiply by a constant, (2) differentiate with respect to time, (3) add or subtract two vectors.

1-6 Newton's laws in vector notation

In order to write Newton's laws in vector form, we have to go just one step further, and define the acceleration vector. This is the time derivative of the velocity vector, and it is easy to demonstrate that its

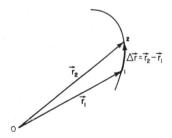

Figure 1-6. The displacement of a particle in a short time interval $\Delta t = t_2 - t_1$.

components are the second derivatives of x, y, and z with respect to t:

$$\mathbf{a} = \frac{d\mathbf{v}}{dt} = \left(\frac{d}{dt}\right)\left(\frac{d\mathbf{r}}{dt}\right) = \frac{d^2\mathbf{r}}{dt^2}, \tag{1.11}$$

$$a_x = \frac{dv_x}{dt} = \frac{d^2x}{dt^2}, \quad a_y = \frac{dv_y}{dt} = \frac{d^2y}{dt^2}, \quad a_z = \frac{dv_z}{dt} = \frac{d^2z}{dt^2}. \tag{1.12}$$

With this definition, then, Newton's laws can be written in this way:

$$m\mathbf{a} = \mathbf{F} \tag{1.13}$$

or

$$m(d^2\mathbf{r}/dt^2) = \mathbf{F}. \tag{1.14}$$

Now the problem of proving the invariance of Newton's laws under rotation of coordinates is this: prove that **a** is a vector; this we have just done. Prove that **F** is a vector; we *suppose* it is. So if force is a vector, then, since we know acceleration is a vector, Eq. (1.13) will look the same in any coordinate system. Writing it in a form which does not explicitly contain x's, y's, and z's has the advantage that from now on we need not write *three* laws every time we write Newton's equations or other laws of physics. We write what looks like *one* law, but really, of course, it is the three laws for any particular set of axes, because any vector equation involves the statement that *each of the components is equal*.

The fact that the acceleration is the rate of change of the vector velocity helps us to calculate the acceleration in some rather complicated circumstances. Suppose, for instance, that a particle is moving on some complicated curve (Figure 1-7) and that, at a given instant t, it had a certain velocity $\mathbf{v}_1$, but that when we go to another instant t_2 a little later, it has a different velocity $\mathbf{v}_2$. What is the acceleration? Answer: Acceleration is the difference in the velocity divided by the small time interval, so we need the difference of the two velocities. How do we get the difference of the velocities? To subtract two vectors, we put the vector across the ends of $\mathbf{v}_2$ and $\mathbf{v}_1$; that is, we

Vectors

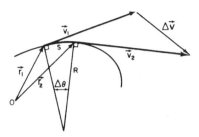

Figure 1-7. A curved trajectory.

draw Δ as the difference of the two vectors, right? *No!* That only works when the *tails* of the vectors are in the same place! It has no meaning if we move the vector somewhere else and then draw a line across, so watch out! We have to draw a new diagram to subtract the vectors. In Figure 1-8, $\mathbf{v}_1$ and $\mathbf{v}_2$ are both drawn parallel and equal to their counterparts in Figure 1-7, and now we can discuss the acceleration. Of course the acceleration is simply $\Delta \mathbf{v}/\Delta t$. It is interesting to note that we can compose the velocity difference out of two parts; we can think of acceleration as having *two components*, $\Delta \mathbf{v}_{\parallel}$ in the direction tangent to the path and $\Delta \mathbf{v}_{\perp}$ at right angles to the path, as indicated in Figure 1-8. The acceleration tangent to the path is, of course, just the change in the *length* of the vector, i.e., the change in the *speed v*:

$$a_{\parallel} = dv/dt. \qquad (1.15)$$

The other component of acceleration, at right angles to the curve, is easy to calculate, using Figures 1-7 and 1-8. In the short time Δt let the change in angle between $\mathbf{v}_1$ and $\mathbf{v}_2$ be the small angle $\Delta \theta$. If

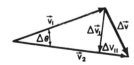

Figure 1-8. Diagram for calculating the acceleration.

the magnitude of the velocity is called v, then of course

$$\Delta v_\perp = v \, \Delta\theta$$

and the acceleration a will be

$$a_\perp = v \, (\Delta\theta/\Delta t).$$

Now we need to know $\Delta\theta/\Delta t$, which can be found this way: If, at the given moment, the curve is approximated as a circle of a certain radius R, then in a time Δt the distance s is, of course, $v \, \Delta t$, where v is the speed.

$$\Delta\theta = v(\Delta t/R), \quad \text{or} \quad \Delta\theta/\Delta t = v/R.$$

Therefore, we find

$$a = v^2/R, \tag{1.16}$$

as we have seen before.

1-7 Scalar product of vectors

Now let us examine a little further the properties of vectors. It is easy to see that the *length* of a step in space would be the same in any coordinate system. That is, if a particular step $\mathbf{r}$ is represented by x, y, z, in one coordinate system, and by x', y', z' in another coordinate system, surely the distance $r = |\mathbf{r}|$ would be the same in both. Now

$$r = \sqrt{x^2 + y^2 + z^2}$$

and also

$$r' = \sqrt{x'^2 + y'^2 + z'^2}.$$

So what we wish to verify is that these two quantities are equal. It is much more convenient not to bother to take the square root, so let us talk about the square of the distance; that is, let us find out whether

$$x^2 + y^2 + z^2 = x'^2 + y'^2 + z'^2. \tag{1.17}$$

It had better be—and if we substitute Eq. (1.5) we do indeed find that it is. So we see that there are other kinds of equations which are true for any two coordinate systems.

Vectors

Something new is involved. We can produce a new quantity, a function of x, y, and z, called a *scalar function*, a quantity which has no direction but which is the same in both systems. Out of a vector we can make a scalar. We have to find a general rule for that. It is clear what the rule is for the case just considered: add the squares of the components. Let us now define a new thing, which we call $\mathbf{a} \cdot \mathbf{a}$. This is not a vector, but a scalar; it is a number that is the same in all coordinate systems, and it is defined to be the sum of the squares of the three components of the vector:

$$\mathbf{a} \cdot \mathbf{a} = a_x^2 + a_y^2 + a_z^2. \tag{1.18}$$

Now you say, "But with what axes?" It does not depend on the axes, the answer is the same in *every* set of axes. So we have a new *kind* of quantity, a new *invariant* or *scalar* produced by one vector "squared." If we now define the following quantity for any two vectors $\mathbf{a}$ and $\mathbf{b}$:

$$\mathbf{a} \cdot \mathbf{b} = a_x b_x + a_y b_y + a_z b_z, \tag{1.19}$$

we find that this quantity, calculated in the primed and unprimed systems, also stays the same. To prove it we note that it is true of $\mathbf{a} \cdot \mathbf{a}$, $\mathbf{b} \cdot \mathbf{b}$, and $\mathbf{c} \cdot \mathbf{c}$, where $\mathbf{c} = \mathbf{a} + \mathbf{b}$. Therefore the sum of the squares $(a_x + b_x)^2 + (a_y + b_y)^2 + (a_z + b_z)^2$ will be invariant:

$$(a_x + b_x)^2 + (a_y + b_y)^2 + (a_z + b_z)^2 =$$
$$(a_{x'} + b_{x'})^2 + (a_{y'} + b_{y'})^2 + (a_{z'} + b_{z'})^2. \tag{1.20}$$

If both sides of this equation are expanded, there will be cross products of just the type appearing in Eq. (1.19), as well as the sums of squares of the components of $\mathbf{a}$ and $\mathbf{b}$. The invariance of terms of the form of Eq. (1.18) then leaves the cross product terms (1.19) invariant also.

The quantity $\mathbf{a} \cdot \mathbf{b}$ is called the *scalar product* of two vectors, $\mathbf{a}$ and $\mathbf{b}$, and it has many interesting and useful properties. For instance, it is easily proved that

$$\mathbf{a} \cdot (\mathbf{b} + \mathbf{c}) = \mathbf{a} \cdot \mathbf{b} + \mathbf{a} \cdot \mathbf{c}. \tag{1.21}$$

Also, there is a simple geometrical way to calculate $\mathbf{a} \cdot \mathbf{b}$, without having to calculate the components of $\mathbf{a}$ and $\mathbf{b}$: $\mathbf{a} \cdot \mathbf{b}$ is the product of the length of $\mathbf{a}$ and the length of $\mathbf{b}$ times the cosine of the angle between them. Why? Suppose that we choose a special coordinate system in which the x-axis lies along $\mathbf{a}$; in those circumstances, the only component of $\mathbf{a}$ that will be there is a_x, which is of course the whole length of $\mathbf{a}$. Thus Eq. (1.19) reduces to $\mathbf{a} \cdot \mathbf{b} = a_x b_x$ for this case, and this is the length of $\mathbf{a}$ times the component of $\mathbf{b}$ in the direction of $\mathbf{a}$, that is, $b \cos \theta$:

$$\mathbf{a} \cdot \mathbf{b} = ab \cos \theta.$$

Therefore, in that special coordinate system, we have proved that $\mathbf{a} \cdot \mathbf{b}$ is the length of $\mathbf{a}$ times the length of $\mathbf{b}$ times $\cos \theta$. But *if it is true in one coordinate system, it is true in all*, because $\mathbf{a} \cdot \mathbf{b}$ is independent of the coordinate system; that is our argument.

What good is the dot product? Are there any cases in physics where we need it? Yes, we need it all the time. For instance, in Chapter 4* the kinetic energy was called $\frac{1}{2}mv^2$, but if the object is moving in space it should be the velocity squared in the x-direction, the y-direction, and the z-direction, and so the formula for kinetic energy according to vector analysis is

$$\text{K.E.} = \tfrac{1}{2}m(\mathbf{v} \cdot \mathbf{v}) = \tfrac{1}{2}m(v_x^2 + v_y^2 + v_z^2). \tag{1.22}$$

Energy does not have direction. Momentum has direction; it is a vector, and it is the mass times the velocity vector.

Another example of a dot product is the work done by a force when something is pushed from one place to the other. We have not yet defined work, but it is equivalent to the energy change, the weights lifted, when a force $\mathbf{F}$ acts through a distance $\mathbf{s}$:

$$\text{Work} = \mathbf{F} \cdot \mathbf{s}. \tag{1.23}$$

It is sometimes very convenient to talk about the component of a vector in a certain direction (say the vertical direction because that is the direction of gravity). For such purposes, it is useful to invent

---

* of the original *Lectures on Physics*, vol. I.

Vectors

what we call a *unit vector* in the direction that we want to study. By a unit vector we mean one whose dot product with itself is equal to unity. Let us call this unit vector $\mathbf{i}$; then $\mathbf{i} \cdot \mathbf{i} = 1$. Then, if we want the component of some vector in the direction of $\mathbf{i}$, we see that the dot product $\mathbf{a} \cdot \mathbf{i}$ will be $a \cos \theta$, i.e., the component of $\mathbf{a}$ in the direction of $\mathbf{i}$. This is a nice way to get the component; in fact, it permits us to get *all* the components and to write a rather amusing formula. Suppose that in a given system of coordinates, x, y, and z, we invent three vectors: $\mathbf{i}$, a unit vector in the direction x; $\mathbf{j}$, a unit vector in the direction y; and $\mathbf{k}$, a unit vector in the direction z. Note first that $\mathbf{i} \cdot \mathbf{i} = 1$. What is $\mathbf{i} \cdot \mathbf{j}$? When two vectors are at right angles, their dot product is zero. Thus

$$\begin{aligned}
\mathbf{i} \cdot \mathbf{i} &= 1 \\
\mathbf{i} \cdot \mathbf{j} &= 0 \quad \mathbf{j} \cdot \mathbf{j} = 1 \\
\mathbf{i} \cdot \mathbf{k} &= 0 \quad \mathbf{j} \cdot \mathbf{k} = 0 \quad \mathbf{k} \cdot \mathbf{k} = 1
\end{aligned} \tag{1.24}$$

Now with these definitions, any vector whatsoever can be written this way:

$$\mathbf{a} = a_x \mathbf{i} + a_y \mathbf{j} + a_z \mathbf{k}. \tag{1.25}$$

By this means we can go from the components of a vector to the vector itself.

This discussion of vectors is by no means complete. However, rather than try to go more deeply into the subject now, we shall first learn to use in physical situations some of the ideas so far discussed. Then, when we have properly mastered this basic material, we shall find it easier to penetrate more deeply into the subject without getting too confused. We shall later find that it is useful to define another kind of product of two vectors, called the vector product, and written as $\mathbf{a} \times \mathbf{b}$. However, we shall undertake a discussion of such matters in a later chapter.

Two

SYMMETRY IN PHYSICAL LAWS

2-1 Symmetry operations

▷ The subject of this chapter is what we may call *symmetry in physical laws*. We have already discussed certain features of symmetry in physical laws in connection with vector analysis (Chapter 1), the theory of relativity (which follows in Chapter 4), and rotation (Chapter 20*).

Why should we be concerned with symmetry? In the first place, symmetry is fascinating to the human mind, and everyone likes objects or patterns that are in some way symmetrical. It is an interesting fact that nature often exhibits certain kinds of symmetry in the objects we find in the world around us. Perhaps the most symmetrical object imaginable is a sphere, and nature is full of spheres—stars, planets, water droplets in clouds. The crystals found in rocks exhibit many different kinds of symmetry, the study of which tells us some important things about the structure of solids. Even the animal and vegetable worlds show some degree of symmetry, although the symmetry of a flower or of a bee is not as perfect or as fundamental as is that of a crystal.

But our main concern here is not with the fact that the *objects* of

---

* of the original *Lectures on Physics*, vol. I.

nature are often symmetrical. Rather, we wish to examine some of the even more remarkable symmetries of the universe—the symmetries that exist in the *basic laws themselves* which govern the operation of the physical world.

First, what *is* symmetry? How can a physical *law* be "symmetrical"? The problem of defining symmetry is an interesting one and we have already noted that Weyl gave a good definition, the substance of which is that a thing is symmetrical if there is something we can do to it so that after we have done it, it looks the same as it did before. For example, a symmetrical vase is of such a kind that if we reflect or turn it, it will look the same as it did before. The question we wish to consider here is what we can do to physical phenomena, or to a physical situation in an experiment, and yet leave the result the same. A list of the known operations under which various physical phenomena remain invariant is shown in Table 2-1.

2-2 *Symmetry in space and time*

The first thing we might try to do, for example, is to *translate* the phenomenon in space. If we do an experiment in a certain region, and then build another apparatus at another place in space (or move the original one over) then, whatever went on in one apparatus, in a certain order in time, will occur in the same way if we

Table 2-1. Symmetry Operations

Translation in space
Translation in time
Rotation through a fixed angle
Uniform velocity in a straight line (Lorentz transformation)
Reversal of time
Reflection of space
Interchange of identical atoms or identical particles
Quantum-mechanical phase
Matter-antimatter (charge conjugation)

have arranged the same condition, with all due attention to the restrictions that we mentioned before: that all of those features of the environment which make it not behave the same way have also been moved over—we talked about how to define how much we should include in those circumstances, and we shall not go into those details again.

In the same way, we also believe today that *displacement in time* will have no effect on physical laws. (That is, *as far as we know today*—all of these things are as far as we know today!) That means that if we build a certain apparatus and start it at a certain time, say on Thursday at 10:00 A.M., and then build the same apparatus and start it, say, three days later in the same condition, the two apparatuses will go through the same motions in exactly the same way as a function of time no matter what the starting time, provided again, of course, that the relevant features of the environment are also modified appropriately in *time*. That symmetry means, of course, that if one bought General Motors stock three months ago, the same thing would happen to it if he bought it now!

We have to watch out for geographical differences too, for there are, of course, variations in the characteristics of the earth's surface. So, for example, if we measure the magnetic field in a certain region and move the apparatus to some other region, it may not work in precisely the same way because the magnetic field is different, but we say that is because the magnetic field is associated with the earth. We can imagine that if we move the whole earth and the equipment, it would make no difference in the operation of the apparatus.

Another thing that we discussed in considerable detail was rotation in space: if we turn an apparatus at an angle it works just as well, provided we turn everything else that is relevant along with it. In fact, we discussed the problem of symmetry under rotation in space in some detail in Chapter 1, and we invented a mathematical system called *vector analysis* to handle it as neatly as possible.

On a more advanced level we had another symmetry—the symmetry under uniform velocity in a straight line. That is to say—a rather remarkable effect—that if we have a piece of apparatus working a certain way and then take the same apparatus and put it in a car, and move the whole car, plus all the relevant surroundings, at a uniform velocity in a straight line, then so far as the phenomena inside the car are concerned there is no difference: all the laws of physics appear the same. We even know how to express this more technically, and that is that the mathematical equations of the physical laws must be unchanged under a *Lorentz transformation*. As a matter of fact, it was a study of the relativity problem that concentrated physicists' attention most sharply on symmetry in physical laws.

Now the above-mentioned symmetries have all been of a geometrical nature, time and space being more or less the same, but there are other symmetries of a different kind. For example, there is a symmetry which describes the fact that we can replace one atom by another of the same kind; to put it differently, there *are* atoms of the same kind. It is possible to find groups of atoms such that if we change a pair around, it makes no difference—the atoms are identical. Whatever one atom of oxygen of a certain type will do, another atom of oxygen of that type will do. One may say, "That is ridiculous, that is the *definition* of equal types!" That may be merely the definition, but then we still do not know whether there *are* any "atoms of the same type"; the *fact* is that there are many, many atoms of the same type. Thus it does mean something to say that it makes no difference if we replace one atom by another of the same type. The so-called elementary particles of which the atoms are made are also identical particles in the above sense—all electrons are the same; all protons are the same; all positive pions are the same; and so on.

After such a long list of things that can be done without changing the phenomena, one might think we could do practically anything; so let us give some examples to the contrary, just to see the difference. Suppose that we ask: "Are the physical laws symmetrical under a change of scale?" Suppose we build a certain piece of apparatus, and then build another apparatus five times bigger in every part, will it work exactly the same way? The answer is, in this case, *no*! The wavelength of light emitted, for example, by the atoms inside one box of sodium atoms and the wavelength of light emitted by a gas of sodium atoms five times in volume is not five times longer, but is in fact exactly the same as the other. So the ratio of the wavelength to the size of the emitter will change.

Another example: we see in the newspaper, every once in a while, pictures of a great cathedral made with little matchsticks—a tremendous work of art by some retired fellow who keeps gluing matchsticks together. It is much more elaborate and wonderful than any real cathedral. If we imagine that this wooden cathedral were actually built on the scale of a real cathedral, we see where the trouble is; it would not last—the whole thing would collapse because of the fact that scaled-up matchsticks are just not strong enough. "Yes," one might say, "but we also know that when there is an influence from the outside, it also must be changed in proportion!" We are talking about the ability of the object to withstand gravitation. So what we should do is first to take the model cathedral of real matchsticks and the real earth, and then we know it is stable. Then we should take the larger cathedral and take a bigger earth. But then it is even worse, because the gravitation is increased still more!

Today, of course, we understand the fact that phenomena depend on the scale on the grounds that matter is atomic in nature, and certainly if we built an apparatus that was so small there were only five atoms in it, it would clearly be something we could not scale up and down arbitrarily. The scale of an individual atom is not at all arbitrary—it is quite definite.

The fact that the laws of physics are not unchanged under a change of scale was discovered by Galileo. He realized that the strengths of materials were not in exactly the right proportion to their sizes, and he illustrated this property that we were just discussing, about the cathedral of matchsticks, by drawing two bones, the bone of one dog, in the right proportion for holding up his weight, and the imaginary bone of a "super dog" that would be, say, ten or a hundred times bigger—that bone was a big, solid thing with quite different proportions. We do not know whether he ever carried the argument quite to the conclusion that the laws of nature must have a definite scale, but he was so impressed with this discovery that he considered it to be as important as the discovery of the laws of motion, because he published them both in the same volume, called "On Two New Sciences."

Another example in which the laws are not symmetrical, that we know quite well, is this: a system in rotation at a uniform angular velocity does not give the same apparent laws as one that is not rotating. If we make an experiment and then put everything in a space ship and have the space ship spinning in empty space, all alone at a constant angular velocity, the apparatus will not work the same way because, as we know, things inside the equipment will be thrown to the outside, and so on, by the centrifugal or coriolis forces, etc. In fact, we can tell that the earth is rotating by using a so-called Foucault pendulum, without looking outside.

Next we mention a very interesting symmetry which is obviously false, i.e., *reversibility in time*. The physical laws apparently cannot be reversible in time, because, as we know, all obvious phenomena are irreversible on a large scale: "The moving finger writes, and having writ, moves on." So far as we can tell, this irreversibility is due to the very large number of particles involved, and if we could see the individual molecules, we would not be able to discern whether the machinery was working forwards or backwards. To make it more precise: we build a small apparatus in which we know what all the atoms are doing, in which we can watch them jiggling. Now we build another apparatus like it, but which starts its motion in the final condition of the other one, with all the velocities precisely reversed. *It will then go through the same motions, but exactly in reverse.* Putting it another way: if we take a motion picture, with sufficient detail, of all the inner works of a piece of material and shine it on a screen and run it backwards, no physicist will be able to say, "That is against the laws of physics, that is doing something wrong!" If we do not see all the details, of course, the situation will be perfectly clear. If we see the egg splattering on the sidewalk and the shell cracking open, and so on, then we will surely say, "That is irreversible, because if we run the moving picture backwards the egg will all collect together and the shell will go back together, and that is obviously ridiculous!" But if we look at the individual atoms themselves, the laws look completely reversible. This is, of course, a much harder discovery to have made, but apparently it is true that the fundamental physical laws, on a microscopic and fundamental level, are completely reversible in time!

2-3 *Symmetry and conservation laws*

The symmetries of the physical laws are very interesting at this level, but they turn out, in the end, to be even more interesting and exciting when we come to quantum mechanics. For a reason which we cannot make clear at the level of the present discussion—a fact that most physicists still find somewhat staggering, a most profound and beautiful thing, is that, in quantum mechanics, *for each of the rules of symmetry there is a corresponding conservation law*; there is a definite connection between the laws of conservation and the symmetries of physical laws. We can only state this at present, without any attempt at explanation.

The fact, for example, that the laws are symmetrical for translation in space when we add the principles of quantum mechanics, turns out to mean that *momentum is conserved*.

That the laws are symmetrical under translation in time means, in quantum mechanics, that *energy is conserved*.

Invariance under rotation through a fixed angle in space corresponds to the *conservation of angular momentum*. These connections are very interesting and beautiful things, among the most beautiful and profound things in physics.

Incidentally, there are a number of symmetries which appear in quantum mechanics which have no classical analog, which have no method of description in classical physics. One of these is as follows: If ψ is the amplitude for some process or other, we know that the absolute square of ψ is the probability that the process will occur. Now if someone else were to make his calculations, not with this ψ, but with a ψ' which differs merely by a change in phase (let Δ be some constant, and multiply $e^{i\Delta}$ times the old ψ), the absolute square of ψ', which is the probability of the event, is then equal to the absolute square of ψ:

$$\psi' = \psi e^{i\Delta}; \qquad |\psi'|^2 = |\psi|^2. \tag{2.1}$$

Therefore the physical laws are unchanged if the phase of the wave function is shifted by an arbitrary constant. That is another symmetry. Physical laws must be of such a nature that a shift in the quantum-mechanical phase makes no difference. As we have just mentioned, in quantum mechanics there is a conservation law for every symmetry. The conservation law which is connected with the quantum-mechanical phase seems to be the *conservation of electrical charge*. This is altogether a very interesting business!

2-4 Mirror reflections

Now the next question, which is going to concern us for most of the rest of this chapter, is the question of symmetry under *reflection in space*. The problem is this: Are the physical laws symmetrical under reflection? We may put it this way: Suppose we build a piece of equipment, let us say a clock, with lots of wheels and hands and numbers; it ticks, it works, and it has things wound up inside. We look at the clock in the mirror. How it *looks* in the mirror is not the question. But let us actually *build* another clock which is exactly the same as the way the first clock looks in the mirror—every time there is a screw with a right-hand thread in one, we use a screw with a left-hand thread in the corresponding place of the other; where one is marked "2" on the face, we mark a "ς" on the face of the other; each coiled spring is twisted one way in one clock and the other way in the mirror-image clock; when we are all finished, we have two clocks, both physical, which bear to each other the relation of an object and its mirror image, although they are both actual, material objects, we emphasize. Now the question is: If the two clocks are

Symmetry in Physical Laws

started in the same condition, the springs wound to corresponding tightnesses, will the two clocks tick and go around, forever after, as exact mirror images? (This is a physical question, not a philosophical question.) Our intuition about the laws of physics would suggest that they *would*.

We would suspect that, at least in the case of these clocks, reflection in space is one of the symmetries of physical laws, that if we change everything from "right" to "left" and leave it otherwise the same, we cannot tell the difference. Let us, then, suppose for a moment that this is true. If it is true, then it would be impossible to distinguish "right" and "left" by any physical phenomenon, just as it is, for example, impossible to define a particular absolute velocity by a physical phenomenon. So it should be impossible, by any physical phenomenon, to define absolutely what we mean by "right" as opposed to "left," because the physical laws should be symmetrical.

Of course, the world does not *have* to be symmetrical. For example, using what we may call "geography," surely "right" can be defined. For instance, we stand in New Orleans and look at Chicago, and Florida is to our right (when our feet are on the ground!). So we can define "right" and "left" by geography. Of course, the actual situation in any system does not have to have the symmetry that we are talking about; it is a question of whether the *laws* are symmetrical—in other words, whether it is *against the physical laws* to have a sphere like the earth with "left-handed dirt" on it and a person like ourselves standing looking at a city like Chicago from a place like New Orleans, but with everything the other way around, so Florida is on the other side. It clearly seems not impossible, not against the physical laws, to have everything changed left for right.

Another point is that our definition of "right" should not depend on history. An easy way to distinguish right from left is to go to a machine shop and pick up a screw at random. The odds are it has a right-hand thread—not necessarily, but it is much more likely to have a right-hand thread than a left-hand one. This is a question of history or convention, or the way things happen to be, and is again not a question of fundamental laws. As we can well appreciate, everyone could have started out making left-handed screws!

So we must try to find some phenomenon in which "right hand" is involved fundamentally. The next possibility we discuss is the fact that polarized light rotates its plane of polarization as it goes through, say, sugar water. As we saw in Chapter 33,* it rotates, let us say, to the right in a certain sugar solution. That is a way of defining "right-hand," because we may dissolve some sugar in the water and then the polarization goes to the right. But sugar has come from living things, and if we try to make the sugar artificially, then we discover that it *does not* rotate the plane of polarization! But if we then take that same sugar which is made artificially and which does not rotate the plane of polarization, and put bacteria in it (they eat some of the sugar) and then filter out the bacteria, we find that we still have sugar left (almost half as much as we had before), and this time it does rotate the plane of polarization, but *the other way*! It seems very confusing, but is easily explained.

Take another example: One of the substances which is common to all living creatures and that is fundamental to life is protein. Proteins consist of chains of amino acids. Figure 2-1 shows a model of an amino acid that comes out of a protein. This amino acid is called alanine, and the molecular arrangement would look like that in Figure 2-1(a) if it came out of a protein of a real living thing. On the other hand, if we try to make alanine from carbon dioxide, ethane, and ammonia (and we *can* make it, it is not a complicated molecule), we discover that we are making equal amounts of this molecule and the one shown in Figure 2-1(b)! The first molecule, the one that comes from the living thing, is called *L-alanine*. The other one, which is the same chemically, in that it has the same kinds of atoms and the same connections of the atoms, is a "right-hand" molecule, compared with the "left-hand" L-alanine, and it is called *D-alanine*. The interesting thing is that when we make alanine at

---

* of the original *Lectures on Physics*, vol. I.

Symmetry in Physical Laws

Figure 2-1. (a) L-alanine (left), and (b) D-alanine (right).

home in a laboratory from simple gases, we get an equal mixture of both kinds. However, the only thing that life uses is L-alanine. (This is not exactly true. Here and there in living creatures there is a special use for D-alanine, but it is very rare. All proteins use L-alanine exclusively.) Now if we make both kinds, and we feed the mixture to some animal which likes to "eat," or use up, alanine, it cannot use D-alanine, so it only uses the L-alanine; that is what happened to our sugar—after the bacteria eat the sugar that works well for them, only the "wrong" kind is left! (Left-handed sugar tastes sweet, but not the same as right-handed sugar.)

So it looks as though the phenomena of life permit a distinction between "right" and "left," or chemistry permits a distinction, because the two molecules are chemically different. But no, it does not! So far as physical measurements can be made, such as of energy, the rates of chemical reactions, and so on, the two kinds work exactly the same way if we make everything else in a mirror image too. One molecule will rotate light to the right, and the other will rotate it to the left in precisely the same amount, through the same amount of fluid. Thus, so far as physics is concerned, these two amino acids are equally satisfactory. So far as we understand things today, the fundamentals of the Schrödinger equation have it that the two molecules should behave in exactly corresponding ways, so that one is to the right as the other is to the left. Nevertheless, in life it is all one way!

It is presumed that the reason for this is the following. Let us suppose, for example, that life is somehow at one moment in a certain condition in which all the proteins in some creatures have left-handed amino acids, and all the enzymes are lopsided—every substance in the living creature is lopsided—it is not symmetrical. So when the digestive enzymes try to change the chemicals in the food from one kind to another, one kind of chemical "fits" into the enzyme, but the other kind does not (like Cinderella and the slipper, except that it is a "left foot" that we are testing). So far as we know, in principle, we could build a frog, for example, in which every molecule is reversed, everything is like the "left-hand" mirror image of a real frog; we have a left-hand frog. This left-hand frog would go on all right for a while, but he would find nothing to eat, because if he swallows a fly, his enzymes are not built to digest it. The fly has the wrong "kind" of amino acids (unless we give him a left-hand fly). So as far as we know, the chemical and life processes would continue in the same manner if everything were reversed.

If life is entirely a physical and chemical phenomenon, then we can understand that the proteins are all made in the same corkscrew only from the idea that at the very beginning some living molecules, by accident, got started and a few won. Somewhere, once, one organic molecule was lopsided in a certain way, and from this particular thing the "right" happened to evolve in our particular geography; a particular historical accident was one-sided, and ever since then the lopsidedness has propagated itself. Once having arrived at the state that it is in now, of course, it will always continue—all the enzymes digest the right things, manufacture the right things: when the carbon dioxide and the water vapor, and so on, go in the plant leaves, the enzymes that make the sugars make them lopsided because the enzymes are lopsided. If any new kind of virus or living thing were to originate at a later time, it would survive only if it could "eat" the kind of living matter already present. Thus it, too, must be of the same kind.

Symmetry in Physical Laws

There is no conservation of the number of right-handed molecules. Once started, we could keep increasing the number of right-handed molecules. So the presumption is, then, that the phenomena in the case of life do not show a lack of symmetry in physical laws, but do show, on the contrary, the universal nature and the commonness of ultimate origin of all creatures on earth, in the sense described above.

2-5 Polar and axial vectors

Now we go further. We observe that in physics there are a lot of other places where we have "right-hand" and "left-hand" rules. As a matter of fact, when we learned about vector analysis we learned about the right-hand rules we have to use in order to get the angular momentum, torque, magnetic field, and so on, to come out right. The force on a charge moving in a magnetic field, for example, is $\mathbf{F} = q\mathbf{v} \times \mathbf{B}$. In a given situation, in which we know $\mathbf{F}$, $\mathbf{v}$, and $\mathbf{B}$, isn't that equation enough to define right-handedness? As a matter of fact, if we go back and look at where the vectors came from, we know that the "right-hand rule" was merely a convention; it was a trick. The original quantities, like the angular momenta and the angular velocities, and things of this kind, were not really vectors at all! They are all somehow associated with a certain plane, and it is just because there are three dimensions in space that we can associate the quantity with a direction perpendicular to that plane. Of the two possible directions, we chose the "right-hand" direction.

So if the laws of physics are symmetrical, we should find that if some demon were to sneak into all the physics laboratories and replace the word "right" for "left" in every book in which "right-hand rules" are given, and instead we were to use all "left-hand rules," uniformly, then it should make no difference whatever in the physical laws.

Let us give an illustration. There are two kinds of vectors. There are "honest" vectors, for example a step $\Delta\mathbf{r}$ in space. If in our apparatus there is a piece here and something else there, then in a mirror apparatus there will be the image piece and the image something else, and if we draw a vector from the "piece" to the "something else," one vector is the mirror image of the other (Figure 2-2). The vector arrow changes its head, just as the whole space turns inside out; such a vector we call a *polar vector*.

But the other kind of vector, which has to do with rotations, is of a different nature. For example, suppose that in three dimensions something is rotating as shown in Figure 2-3. Then if we look at it in a mirror, it will be rotating as indicated, namely, as the mirror image of the original rotation. Now we have agreed to represent the mirror rotation by the same rule, it is a "vector" which, on reflection, does *not* change about as the polar vector does, but is reversed relative to the polar vectors and to the geometry of the space; such a vector is called an *axial vector*.

Now if the law of reflection symmetry is right in physics, then it must be true that the equations must be so designed that if we change the sign of each axial vector and each cross-product of vectors, which would be what corresponds to reflection, nothing will happen. For instance, when we write a formula which says that the angular momentum is $\mathbf{L} = \mathbf{r} \times \mathbf{p}$, that equation is all right, because if we change to a left-hand coordinate system, we change the sign of $\mathbf{L}$, but $\mathbf{p}$ and $\mathbf{r}$ do not change; the cross-product sign is changed, since we must change from a right-hand rule to a left-hand rule. As another example, we know that the force on a charge moving in a magnetic field is $\mathbf{F} = q\mathbf{v} \times \mathbf{B}$, but if we change from a right- to a left-handed system, since $\mathbf{F}$ and $\mathbf{v}$ are known to be polar vectors the sign change required by the cross-product must be cancelled by a sign change in $\mathbf{B}$, which means that $\mathbf{B}$ must be an

Figure 2-2. A step in space and its mirror image.

Symmetry in Physical Laws

Figure 2-3. A rotating wheel and its mirror image. Note that the angular velocity "vector" is not reversed in direction.

axial vector. In other words, if we make such a reflection, **B** must go to − **B**. So if we change our coordinates from right to left, we must also change the poles of magnets from north to south.

Let us see how that works in an example. Suppose that we have two magnets, as in Figure 2-4. One is a magnet with the coils going around a certain way, and with current in a given direction. The other magnet looks like the reflection of the first magnet in a mirror—the coil will wind the other way, everything that happens inside the coil is exactly reversed, and the current goes as shown. Now, from the laws for the production of magnetic fields, which we do not know yet officially, but which we most likely learned in high school, it turns out that the magnetic field is as shown in the figure. In one case the pole is a south magnetic pole, while in the other magnet the current is going the other way and the magnetic field is reversed—it is a north magnetic pole. So we see that when we go from right to left we must indeed change from north to south!

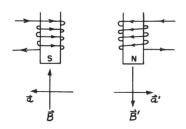

Figure 2-4. A magnet and its mirror image.

Never mind changing north to south; these too are mere conventions. Let us talk about *phenomena*. Suppose, now, that we have an electron moving through one field, going into the page. Then, if we use the formula for the force, $\mathbf{v} \times \mathbf{B}$ (remember the charge is minus), we find that the electron will deviate in the indicated direction according to the physical law. So the phenomenon is that we have a coil with a current going in a specified sense and an electron curves in a certain way—that is the physics—never mind how we label everything.

Now let us do the same experiment with a mirror: we send an electron through in a corresponding direction and now the force is reversed, if we calculate it from the same rule, and that is very good because the corresponding *motions* are then mirror images!

2-6 Which hand is right?

So the fact of the matter is that in studying any phenomenon there are always two right-hand rules, or an even number of them, and the net result is that the phenomena always look symmetrical. In short, therefore, we cannot tell right from left if we also are not able to tell north from south. However, it may seem that we *can* tell the north pole of a magnet. The north pole of a compass needle, for example, is one that points to the north. But of course that is again a local property that has to do with geography of the earth; that is just like talking about in which direction is Chicago, so it does not count. If we have seen compass needles, we may have noticed that the north-seeking pole is a sort of bluish color. But that is just due to the man who painted the magnet. These are all local, conventional criteria.

However, if a magnet were to have the property that if we looked at it closely enough we would see small hairs growing on its north pole but not on its south pole, if that were the general rule, or if there were *any* unique way to distinguish the north from the south pole of a magnet, then we could tell which of the two cases we actually had, and *that would be the end of the law of reflection symmetry*.

Symmetry in Physical Laws

To illustrate the whole problem still more clearly, imagine that we were talking to a Martian, or someone very far away, by telephone. We are not allowed to send him any actual samples to inspect; for instance, if we could send light, we could send him right-hand circularly polarized light and say, "That is right-hand light—just watch the way it is going." But we cannot *give* him anything, we can only talk to him. He is far away, or in some strange location, and he cannot see anything we can see. For instance, we cannot say, "Look at Ursa major; now see how those stars are arranged. What we mean by 'right' is . . ." We are only allowed to telephone him.

Now we want to tell him all about us. Of course, first we start defining numbers, and say, "Tick, tick, *two*, tick, tick, tick, *three* . . . ," so that gradually he can understand a couple of words, and so on. After a while we may become very familiar with this fellow, and he says, "What do you guys look like?" We start to describe ourselves, and say, "Well, we are six feet tall." He says, "Wait a minute, what is six feet?" Is it possible to tell him what six feet is? Certainly! We say, "You know about the diameter of hydrogen atoms—we are 17,000,000,000 hydrogen atoms high!" That is possible because physical laws are not variant under change of scale, and therefore we *can* define an absolute length. And so we define the size of the body, and tell him what the general shape is—it has prongs with five bumps sticking out on the ends, and so on, and he follows us along, and we finish describing how we look on the outside, presumably without encountering any particular difficulties. He is even making a model of us as we go along. He says, "My, you are certainly very handsome fellows; now what is on the inside?" So we start to describe the various organs on the inside, and we come to the heart, and we carefully describe the shape of it, and say, "Now put the heart on the left side." He says, "Duhhh—the left side?" Now our problem is to describe to him which side the heart goes on without his ever seeing anything that we see, and without our ever sending any sample to him of what we mean by "right"—no standard right-handed object. Can we do it?

2-7 *Parity is not conserved!*

It turns out that the laws of gravitation, the laws of electricity and magnetism, nuclear forces, all satisfy the principle of reflection symmetry, so these laws, or anything derived from them, cannot be used. But associated with the many particles that are found in nature there is a phenomenon called *beta decay*, or *weak decay*. One of the examples of weak decay, in connection with a particle discovered in about 1954, posed a strange puzzle. There was a certain charged particle which disintegrated into three π-mesons, as shown schematically in Figure 2-5. This particle was called, for a while, a τ-meson. Now in Figure 2-5 we also see another particle which disintegrates into *two* mesons; one must be neutral, from the conservation of charge. This particle was called a θ-meson. So on the one hand we have a particle called a τ, which disintegrates into three π-mesons, and a θ, which disintegrates into two π-mesons. Now it was soon discovered that the τ and the θ are almost equal in mass; in fact, within the experimental error, they are equal. Next, the length of time it took for them to disintegrate into three π's and two π's was found to be almost exactly the same; they live the same length of time. Next, whenever they were made, they were made in the same proportions, say, 14 percent τ's to 86 percent θ's.

Anyone in his right mind realizes immediately that they must be the same particle, that we merely produce an object which has two different ways of disintegrating—not two different particles. This object that can disintegrate in two different ways has, therefore,

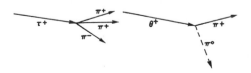

Figure 2-5. A schematic diagram of the disintegration of a τ^+ and a θ^+ particle.

the same lifetime and the same production ratio (because this is simply the ratio of the odds with which it disintegrates into these two kinds).

However, it was possible to prove (and we cannot here explain at all *how*), from the principle of reflection symmetry in quantum mechanics, that it was *impossible* to have these both come from the same particle—the same particle *could not* disintegrate in both of these ways. The conservation law corresponding to the principle of reflection symmetry is something which has no classical analog, and so this kind of quantum-mechanical conservation was called the *conservation of parity*. So, it was a result of the conservation of parity or, more precisely, from the symmetry of the quantum-mechanical equations of the weak decays under reflection, that the same particle could not go into both, so it must be some kind of coincidence of masses, lifetimes, and so on. But the more it was studied, the more remarkable the coincidence, and the suspicion gradually grew that possibly the deep law of the reflection symmetry of nature may be false.

As a result of this apparent failure, the physicists Lee and Yang suggested that other experiments be done in related decays to try to test whether the law was correct in other cases. The first such experiment was carried out by Miss Wu from Columbia, and was done as follows. Using a very strong magnet at a very low temperature, it turns out that a certain isotope of cobalt, which disintegrates by emitting an electron, is magnetic, and if the temperature is low enough that the thermal oscillations do not jiggle the atomic magnets about too much, they line up in the magnetic field. So the cobalt atoms will all line up in this strong field. They then disintegrate, emitting an electron, and it was discovered that when the atoms were lined up in a field whose **B** vector points upward, most of the electrons were emitted in a downward direction.

If one is not really "hep" to the world, such a remark does not sound like anything of significance, but if one appreciates the problems and interesting things in the world, then he sees that it is a most dramatic discovery: When we put cobalt atoms in an extremely strong magnetic field, more disintegration electrons go down than up. Therefore if we were to put it in a corresponding experiment in a "mirror," in which the cobalt atoms would be lined up in the opposite direction, they would spit their electrons *up*, not *down*; the action is *unsymmetrical. The magnet has grown hairs!* The south pole of a magnet is of such a kind that the electrons in a β-disintegration tend to go away from it; that distinguishes, in a physical way, the north pole from the south pole.

After this, a lot of other experiments were done: the disintegration of the π into μ and v; μ into an electron and two neutrinos; nowadays, the Λ into proton and π; disintegration of Σ's; and many other disintegrations. In fact, in almost all cases where it could be expected, all have been found *not* to obey reflection symmetry! Fundamentally, the law of reflection symmetry, at this level in physics, is incorrect.

In short, we can tell a Martian where to put the heart: we say, "Listen, build yourself a magnet, and put the coils in, and put the current on, and then take some cobalt and lower the temperature. Arrange the experiment so the electrons go from the foot to the head, then the direction in which the current goes through the coils is the direction that goes in on what we call the right and comes out on the left." So it is possible to define right and left, now, by doing an experiment of this kind.

There are a lot of other features that were predicted. For example, it turns out that the spin, the angular momentum, of the cobalt nucleus before disintegration is 5 units of $\hbar$, and after disintegration it is 4 units. The electron carries spin angular momentum, and there is also a neutrino involved. It is easy to see from this that the electron must carry its spin angular momentum aligned along its direction of motion, the neutrino likewise. So it looks as though the electron is spinning to the left, and that was also checked. In fact, it was checked right here at Caltech by Boehm and Wapstra, that the electrons spin mostly to the left.

Symmetry in Physical Laws

(There were some other experiments that gave the opposite answer, but they were wrong!)

The next problem, of course, was to find the law of the failure of parity conservation. What is the rule that tells us how strong the failure is going to be? The rule is this: it occurs only in these very slow reactions, called weak decays, and when it occurs, the rule is that the particles which carry spin, like the electron, neutrino, and so on, come out with a spin tending to the left. That is a lopsided rule; it connects a polar vector velocity and an axial vector angular momentum, and says that the angular momentum is more likely to be opposite to the velocity than along it.

Now that is the rule, but today we do not really understand the whys and wherefores of it. *Why* is this the right rule, what is the fundamental reason for it, and how is it connected to anything else? At the moment we have been so shocked by the fact that this thing is unsymmetrical that we have not been able to recover enough to understand what it means with regard to all the other rules. However, the subject is interesting, modern, and still unsolved, so it seems appropriate that we discuss some of the questions associated with it.

2-8 Antimatter

The first thing to do when one of the symmetries is lost is to immediately go back over the list of known or assumed symmetries and ask whether any of the others are lost. Now we did not mention one operation on our list, which must necessarily be questioned, and that is the relation between matter and antimatter. Dirac predicted that in addition to electrons there must be another particle, called the positron (discovered at Caltech by Anderson), that is necessarily related to the electron. All the properties of these two particles obey certain rules of correspondence: the energies are equal; the masses are equal; the charges are reversed; but, more important than anything, the two of them, when they come together, can annihilate each other and liberate their entire mass in the form of energy, say γ-rays. The positron is called an *antiparticle* to the electron, and these are the characteristics of a particle and its antiparticle. It was clear from Dirac's argument that all the rest of the particles in the world should also have corresponding antiparticles. For instance, for the proton there should be an antiproton, which is now symbolized by a $\bar{p}$. The $\bar{p}$ would have a negative electrical charge and the same mass as a proton, and so on. The most important feature, however, is that a proton and an antiproton coming together can annihilate each other. The reason we emphasize this is that people do not understand it when we say there is a neutron and also an antineutron, because they say, "A neutron is neutral, so how *can* it have the opposite charge?" The rule of the "anti" is not just that it has the opposite charge, it has a certain set of properties, the whole lot of which are opposite. The antineutron is distinguished from the neutron in this way: if we bring two neutrons together, they just stay as two neutrons, but if we bring a neutron and an antineutron together, they annihilate each other with a great explosion of energy being liberated, with various π-mesons, γ-rays, and whatnot.

Now if we have antineutrons, antiprotons, and antielectrons, we can make antiatoms, in principle. They have not been made yet, but it is possible in principle. For instance, a hydrogen atom has a proton in the center with an electron going around outside. Now imagine that somewhere we can make an antiproton with a positron going around, would it go around? Well, first of all, the antiproton is electrically negative and the antielectron is electrically positive, so they attract each other in a corresponding manner—the masses are all the same; everything is the same. It is one of the principles of the symmetry of physics, the equations seem to show, that if a clock, say, were made of matter on one hand, and then we made the same clock of antimatter, it would run in this way. (Of course, if we put the clocks together, they would annihilate each other, but that is different.)

An immediate question then arises. We can build, out of matter, two clocks, one which is "left-hand" and one which is "right-hand." For example, we could build a clock which is not built in a simple way, but has cobalt and magnets and electron detectors which detect the presence of β-decay electrons and count them. Each time one is counted, the second hand moves over. Then the mirror clock, receiving fewer electrons, will not run at the same rate. So evidently we can make two clocks such that the left-hand clock does not agree with the right-hand one. Let us make, out of matter, a clock which we call the standard or right-hand clock. Now let us make, also out of matter, a clock which we call the left-hand clock. We have just discovered that, in general, these two will *not* run the same way; prior to that famous physical discovery, it was thought that they would. Now it was also supposed that matter and antimatter were equivalent. That is, if we made an antimatter clock, right-hand, the same shape, then it would run the same as the right-hand matter clock, and if we made the same clock to the left it would run the same. In other words, in the beginning it was believed that *all four* of these clocks were the same; now of course we know that the right-hand and left-hand matter are not the same. Presumably, therefore, the right-handed antimatter and the left-handed antimatter are not the same.

So the obvious question is, which goes with which, if either? In other words, does the right-handed matter behave the same way as the right-handed antimatter? Or does the right-handed matter behave the same way as the left-handed antimatter? β-decay experiments, using positron decay instead of electron decay, indicate that this is the interconnection: matter to the "right" works the same way as antimatter to the "left."

Therefore, at long last, it is really true that right and left symmetry is still maintained! If we made a left-hand clock, but made it out of the other kind of matter, antimatter instead of matter, it would run in the same way. So what has happened is that instead of having two independent rules in our list of symmetries, two of these rules go together to make a new rule, which says that matter to the right is symmetrical with antimatter to the left.

So if our Martian is made of antimatter and we give him instructions to make this "right" handed model like us, it will, of course, come out the other way around. What would happen when, after much conversation back and forth, we each have taught the other to make space ships and we meet halfway in empty space? We have instructed each other on our traditions, and so forth, and the two of us come rushing out to shake hands. Well, if he puts out his left hand, watch out!

2-9 Broken symmetries

The next question is, what can we make out of laws which are *nearly* symmetrical? The marvelous thing about it all is that for such a wide range of important, strong phenomena—nuclear forces, electrical phenomena, and even weak ones like gravitation—over a tremendous range of physics, all the laws for these seem to be symmetrical. On the other hand, this little extra piece says, "No, the laws are not symmetrical!" How is it that nature can be almost symmetrical, but not perfectly symmetrical? What shall we make of this? First, do we have any other examples? The answer is, we do, in fact, have a few other examples. For instance, the nuclear part of the force between proton and proton, between neutron and neutron, and between neutron and proton, is all exactly the same—there is a symmetry for nuclear forces, a new one, that we can interchange neutron and proton—but it evidently is not a general symmetry, for the electrical repulsion between two protons at a distance does not exist for neutrons. So it is not generally true that we can *always* replace a proton with a neutron, but only to a good approximation. Why *good*? Because the nuclear forces are much stronger than the electrical forces. So this is an "almost" symmetry also. So we do have examples in other things.

We have, in our minds, a tendency to accept symmetry as some kind of perfection. In fact it is like the old idea of the Greeks that circles were perfect, and it was rather horrible to believe that the planetary orbits were not circles, but only nearly circles. The differ-

Symmetry in Physical Laws

ence between being a circle and being nearly a circle is not a small difference, it is a fundamental change so far as the mind is concerned. There is a sign of perfection and symmetry in a circle that is not there the moment the circle is slightly off—that is the end of it—it is no longer symmetrical. Then the question is why it is only *nearly* a circle—that is a much more difficult question. The actual motion of the planets, in general, should be ellipses, but during the ages, because of tidal forces, and so on, they have been made almost symmetrical. Now the question is whether we have a similar problem here. The problem from the point of view of the circles is if they were perfect circles there would be nothing to explain, that is clearly simple. But since they are only nearly circles, there is a lot to explain, and the result turned out to be a big dynamical problem, and now our problem is to explain why they are nearly symmetrical by looking at tidal forces and so on.

So our problem is to explain where symmetry comes from. Why is nature so nearly symmetrical? No one has any idea why. The only thing we might suggest is something like this: There is a gate in Japan, a gate in Neiko, which is sometimes called by the Japanese the most beautiful gate in all Japan; it was built in a time when there was great influence from Chinese art. This gate is very elaborate, with lots of gables and beautiful carving and lots of columns and dragon heads and princes carved into the pillars, and so on. But when one looks closely he sees that in the elaborate and complex design along one of the pillars, one of the small design elements is carved upside down; otherwise the thing is completely symmetrical. If one asks why this is, the story is that it was carved upside down so that the gods will not be jealous of the perfection of man. So they purposely put an error in there, so that the gods would not be jealous and get angry with human beings.

We might like to turn the idea around and think that the true explanation of the near symmetry of nature is this: that God made the laws only nearly symmetrical so that we should not be jealous of His perfection!

Three

THE SPECIAL THEORY OF RELATIVITY

3-1 The principle of relativity

◆ For over 200 years the equations of motion enunciated by Newton were believed to describe nature correctly, and the first time that an error in these laws was discovered, the way to correct it was also discovered. Both the error and its correction were discovered by Einstein in 1905.

Newton's Second Law, which we have expressed by the equation

$$F = d(mv)/dt,$$

was stated with the tacit assumption that m is a constant, but we now know that this is not true, and that the mass of a body increases with velocity. In Einstein's corrected formula m has the value

$$m = \frac{m_0}{\sqrt{1 - v^2/c^2}}, \qquad (3.1)$$

where the "rest mass" m_0 represents the mass of a body that is not moving and c is the speed of light, which is about $3 \times 10^5 \, \text{km} \cdot \text{sec}^{-1}$ or about $186,000 \, \text{mi} \cdot \text{sec}^{-1}$.

For those who want to learn just enough about it so they can solve problems, that is all there is to the theory of relativity—it just changes Newton's laws by introducing a correction factor to the

49

mass. From the formula itself it is easy to see that this mass increase is very small in ordinary circumstances. If the velocity is even as great as that of a satellite, which goes around the earth at 5 mi/sec, then $v/c = 5/186,000$: putting this value into the formula shows that the correction to the mass is only one part in two to three billion, which is nearly impossible to observe. Actually, the correctness of the formula has been amply confirmed by the observation of many kinds of particles, moving at speeds ranging up to practically the speed of light. However, because the effect is ordinarily so small, it seems remarkable that it was discovered theoretically before it was discovered experimentally. Empirically, at a sufficiently high velocity, the effect is very large, but it was not discovered that way. Therefore it is interesting to see how a law that involved so delicate a modification (at the time when it was first discovered) was brought to light by a combination of experiments and physical reasoning. Contributions to the discovery were made by a number of people, the final result of whose work was Einstein's discovery.

There are really two Einstein theories of relativity. This chapter is concerned with the Special Theory of Relativity, which dates from 1905. In 1915 Einstein published an additional theory, called the General Theory of Relativity. This latter theory deals with the extension of the Special Theory to the case of the law of gravitation; we shall not discuss the General Theory here.

The principle of relativity was first stated by Newton, in one of his corollaries to the laws of motion: "The motions of bodies included in a given space are the same among themselves, whether that space is at rest or moves uniformly forward in a straight line." This means, for example, that if a space ship is drifting along at a uniform speed, all experiments performed in the space ship and all the phenomena in the space ship will appear the same as if the ship were not moving, provided, of course, that one does not look outside. That is the meaning of the principle of relativity. This is a simple enough idea, and the only question is whether it is *true* that in all experiments performed inside a moving system the laws of physics will appear the same as they would if the system were

The Special Theory of Relativity

standing still. Let us first investigate whether Newton's laws appear the same in the moving system.

Suppose that Moe is moving in the x-direction with a uniform velocity u, and he measures the position of a certain point shown in Figure 3-1. He designates the "x-distance" of the point in his coordinate system as x'. Joe is at rest, and measures the position of the same point, designating its x-coordinate in his system as x. The relationship of the coordinates in the two systems is clear from the diagram. After time t Moe's origin has moved a distance ut, and if the two systems originally coincided,

$$\begin{aligned}
x' &= x - ut, \\
y' &= y, \\
z' &= z, \\
t' &= t.
\end{aligned} \tag{3.2}$$

If we substitute this transformation of coordinates into Newton's laws we find that these laws transform to the same laws in the primed system; that is, the laws of Newton are of the same form in a moving system as in a stationary system, and therefore it is impossible to tell, by making mechanical experiments, whether the system is moving or not.

The principle of relativity has been used in mechanics for a long time. It was employed by various people, in particular Huygens, to obtain the rules for the collision of billiard balls, in much the same

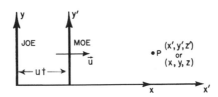

Figure 3-1. Two coordinate systems in uniform relative motion along their x-axes.

way as we used it in Chapter 10* to discuss the conservation of momentum. In the past century interest in it was heightened as the result of investigations into the phenomena of electricity, magnetism, and light. A long series of careful studies of these phenomena by many people culminated in Maxwell's equations of the electromagnetic field, which describe electricity, magnetism, and light in one uniform system. However, the Maxwell equations did *not* seem to obey the principle of relativity. That is, if we transform Maxwell's equations by the substitution of equations 3.2, *their form does not remain the same*; therefore, in a moving space ship the electrical and optical phenomena should be different from those in a stationary ship. Thus one could use these optical phenomena to determine the speed of the ship; in particular, one could determine the absolute speed of the ship by making suitable optical or electrical measurements. One of the consequences of Maxwell's equations is that if there is a disturbance in the field such that light is generated, these electromagnetic waves go out in all directions equally and at the same speed c, or 186,000 mi/sec. Another consequence of the equations is that if the source of the disturbance is moving, the light emitted goes through space at the same speed c. This is analogous to the case of sound, the speed of sound waves being likewise independent of the motion of the source.

This independence of the motion of the source, in the case of light, brings up an interesting problem:

Suppose we are riding in a car that is going at a speed u, and light from the rear is going past the car with speed c. Differentiating the first equation in (3.2) gives

$$dx'/dt = dx/dt - u,$$

which means that according to the Galilean transformation the apparent speed of the passing light, as we measure it in the car, should not be c but should be $c - u$. For instance, if the car is going

---

* of the original *Lectures on Physics* vol. I.

The Special Theory of Relativity

100,000 mi/sec, and the light is going 186,000 mi/sec, then apparently the light going past the car should go 86,000 mi/sec. In any case, by measuring the speed of the light going past the car (if the Galilean transformation is correct for light), one could determine the speed of the car. A number of experiments based on this general idea were performed to determine the velocity of the earth, but they all failed—they gave *no velocity at all*. We shall discuss one of these experiments in detail, to show exactly what was done and what was the matter; something *was* the matter, of course, something was wrong with the equations of physics. What could it be?

3-2 *The Lorentz transformation*

When the failure of the equations of physics in the above case came to light, the first thought that occurred was that the trouble must lie in the new Maxwell equations of electrodynamics, which were only 20 years old at the time. It seemed almost obvious that these equations must be wrong, so the thing to do was to change them in such a way that under the Galilean transformation the principle of relativity would be satisfied. When this was tried, the new terms that had to be put into the equations led to predictions of new electrical phenomena that did not exist at all when tested experimentally, so this attempt had to be abandoned. Then it gradually became apparent that Maxwell's laws of electrodynamics were correct, and the trouble must be sought elsewhere.

In the meantime, H. A. Lorentz noticed a remarkable and curious thing when he made the following substitutions in the Maxwell equations:

$$x' = \frac{x - ut}{\sqrt{1 - u^2/c^2}},$$

$$y' = y,$$

$$z' = z,$$ \hspace{2cm} (3.3)

$$t' = \frac{t - ux/c^2}{\sqrt{1 - u^2/c^2}},$$

namely, Maxwell's equations remain in the same form when this transformation is applied to them! Equations (3.3) are known as a *Lorentz transformation*. Einstein, following a suggestion originally made by Poincaré, then proposed that *all the physical laws* should be of such a kind that they *remain unchanged under a Lorentz transformation*. In other words, we should change, not the laws of electrodynamics, but the laws of mechanics. How shall we change Newton's laws so that *they* will remain unchanged by the Lorentz transformation? If this goal is set, we then have to rewrite Newton's equations in such a way that the conditions we have imposed are satisfied. As it turned out, the only requirement is that the mass m in Newton's equations must be replaced by the form shown in Eq. (3.1). When this change is made, Newton's laws and the laws of electrodynamics will harmonize. Then if we use the Lorentz transformation in comparing Moe's measurements with Joe's, we shall never be able to detect whether either is moving, because the form of all the equations will be the same in both coordinate systems!

It is interesting to discuss what it means that we replace the old transformation between the coordinates and time with a new one, because the old one (Galilean) seems to be self-evident, and the new one (Lorentz) looks peculiar. We wish to know whether it is logically and experimentally possible that the new, and not the old, transformation can be correct. To find that out, it is not enough to study the laws of mechanics but, as Einstein did, we too must analyze our ideas of *space* and *time* in order to understand this transformation. We shall have to discuss these ideas and their implications for mechanics at some length, so we say in advance that the effort will be justified, since the results agree with experiment.

3-3 The Michelson-Morley experiment

As mentioned above, attempts were made to determine the absolute velocity of the earth through the hypothetical "ether" that was supposed to pervade all space. The most famous of these experiments is

The Special Theory of Relativity

one performed by Michelson and Morley in 1887. It was 18 years later before the negative results of the experiment were finally explained, by Einstein.

The Michelson-Morley experiment was performed with an apparatus like that shown schematically in Figure 3-2. This apparatus is essentially comprised of a light source A, a partially silvered glass plate B, and two mirrors C and E, all mounted on a rigid base. The mirrors are placed at equal distances L from B. The plate B splits an oncoming beam of light, and the two resulting beams continue in mutually perpendicular directions to the mirrors, where they are reflected back to B. On arriving back at B, the two beams are recombined as two superposed beams, D and F. If the time taken for the light to go from B to E and back is the same as the time from B to C and back, the emerging beams D and F will be in phase and will reinforce each other, but if the two times differ slightly, the beams will be slightly out of phase and interference will result. If the apparatus is "at rest" in the ether, the times should be precisely equal, but if it is moving toward the right with a velocity u, there should be a difference in the times. Let us see why.

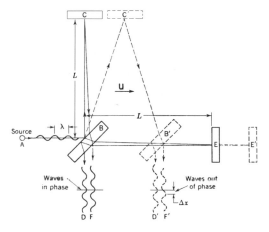

Figure 3-2. Schematic diagram of the Michelson-Morley experiment.

First, let us calculate the time required for the light to go from B to E and back. Let us say that the time for light to go from plate B to mirror E is t_1, and the time for the return is t_2. Now, while the light is on its way from B to the mirror, the apparatus moves a distance ut_1, so the light must traverse a distance $L + ut_1$, at the speed c. We can also express this distance as ct_1, so we have

$$ct_1 = L + ut_1, \quad \text{or} \quad t_1 = L/(c - u).$$

(This result is also obvious from the point of view that the velocity of light relative to the apparatus is $c - u$, so the time is the length L divided by $c - u$.) In a like manner, the time t_2 can be calculated. During this time the plate B advances a distance ut_2, so the return distance of the light is $L - ut_2$. Then we have

$$ct_2 = L - ut_2, \quad \text{or} \quad t_2 = L/(c + u).$$

Then the total time is

$$t_1 + t_2 = 2Lc/(c^2 - u^2).$$

For convenience in later comparison of times we write this as

$$t_1 + t_2 = \frac{2L/c}{1 - u^2/c^2}. \tag{3.4}$$

Our second calculation will be of the time t_3 for the light to go from B to the mirror C. As before, during time t_3 the mirror C moves to the right a distance ut_3 to the position C'; in the same time, the light travels a distance ct_3 along the hypotenuse of a triangle, which is BC'. For this right triangle we have

$$(ct_3)^2 = L^2 + (ut_3)^2$$

or

$$L^2 = c^2 t_3^2 - u^2 t_3^2 = (c^2 - u^2)t_3^2,$$

from which we get

$$t_3 = L/\sqrt{c^2 - u^2}.$$

The Special Theory of Relativity

For the return trip from C' the distance is the same, as can be seen from the symmetry of the figure; therefore the return time is also the same, and the total time is $2t_3$. With a little rearrangement of the form we can write

$$2t_3 = \frac{2L}{\sqrt{c^2 - u^2}} = \frac{2L/c}{\sqrt{1 - u^2/c^2}} \, . \tag{3.5}$$

We are now able to compare the times taken by the two beams of light. In expressions (3.4) and (3.5) the numerators are identical, and represent the time that would be taken if the apparatus were at rest. In the denominators, the term u^2/c^2 will be small, unless u is comparable in size to c. The denominators represent the modifications in the times caused by the motion of the apparatus. And behold, these modifications are *not the same*—the time to go to C and back is a little less than the time to E and back, even though the mirrors are equidistant from B, and all we have to do is to measure that difference with precision.

Here a minor technical point arises—suppose the two lengths L are not exactly equal? In fact, we surely cannot make them exactly equal. In that case we simply turn the apparatus 90 degrees, so that BC is in the line of motion and BE is perpendicular to the motion. Any small difference in length then becomes unimportant, and what we look for is a *shift* in the interference fringes when we rotate the apparatus.

In carrying out the experiment, Michelson and Morley oriented the apparatus so that the line BE was nearly parallel to the earth's motion in its orbit (at certain times of the day and night). This orbital speed is about 18 miles per second, and any "ether drift" should be at least that much at some time of the day or night and at some time during the year. The apparatus was amply sensitive to observe such an effect, but no time difference was found—the velocity of the earth through the ether could not be detected. The result of the experiment was null.

The result of the Michelson-Morley experiment was very

puzzling and most disturbing. The first fruitful idea for finding a way out of the impasse came from Lorentz. He suggested that material bodies contract when they are moving, and that this foreshortening is only in the direction of the motion, and also, that if the length is L_0 when a body is at rest, then when it moves with speed u parallel to its length, the new length, which we call $L_\parallel$ (L-parallel), is given by

$$L_\parallel = L_0\sqrt{1 - u^2/c^2}. \tag{3.6}$$

When this modification is applied to the Michelson-Morley interferometer apparatus the distance from B to C does not change, but the distance from B to E is shortened to $L\sqrt{1 - u^2/c^2}$. Therefore Eq. (3.5) is not changed, but the L of Eq. (3.4) must be changed in accordance with Eq. (3.6). When this is done we obtain

$$t_1 + t_2 = \frac{(2L/c)\sqrt{1 - u^2/c^2}}{1 - u^2/c^2} = \frac{2L/c}{\sqrt{1 - u^2/c^2}}. \tag{3.7}$$

Comparing this result with Eq. (3.5), we see that $t_1 + t_2 = 2t_3$. So if the apparatus shrinks in the manner just described, we have a way of understanding why the Michelson-Morley experiment gives no effect at all. Although the contraction hypothesis successfully accounted for the negative result of the experiment, it was open to the objection that it was invented for the express purpose of explaining away the difficulty, and was too artificial. However, in many other experiments to discover an ether wind, similar difficulties arose, until it appeared that nature was in a "conspiracy" to thwart man by introducing some new phenomenon to undo every phenomenon that he thought would permit a measurement of u.

It was ultimately recognized, as Poincaré pointed out, that *a complete conspiracy is itself a law of nature*! Poincaré then proposed that there *is* such a law of nature, that it is not possible to discover an ether wind by *any* experiment; that is, there is no way to determine an absolute velocity.

The Special Theory of Relativity

3-4 Transformation of time

In checking out whether the contraction idea is in harmony with the facts in other experiments, it turns out that everything is correct provided that the *times* are also modified, in the manner expressed in the fourth equation of the set (3.3). That is because the time t_3, calculated for the trip from B to C and back, is not the same when calculated by a man performing the experiment in a moving space ship as when calculated by a stationary observer who is watching the space ship. To the man in the ship the time is simply $2L/c$, but to the other observer it is $(2L/c)/\sqrt{1 - u^2/c^2}$ (Eq. 3.5). In other words, when the outsider sees the man in the space ship lighting a cigar, all the actions appear to be slower than normal, while to the man inside, everything moves at a normal rate. So not only must the lengths shorten, but also the time-measuring instruments ("clocks") must apparently slow down. That is, when the clock in the space ship records 1 second elapsed, as seen by the man in the ship, it shows $1/\sqrt{1 - u^2/c^2}$ second to the man outside.

This slowing of the clocks in a moving system is a very peculiar phenomenon, and is worth an explanation. In order to understand this, we have to watch the machinery of the clock and see what happens when it is moving. Since that is rather difficult, we shall take a very simple kind of clock. The one we choose is rather a silly kind of clock, but it will work in principle: it is a rod (meter stick) with a mirror at each end, and when we start a light signal between the mirrors, the light keeps going up and down, making a click every time it comes down, like a standard ticking clock. We build two such clocks, with exactly the same lengths, and synchronize them by starting them together; then they agree always thereafter, because they are the same in length, and light always travels with speed c. We give one of these clocks to the man to take along in his space ship, and he mounts the rod perpendicular to the direction of motion of the ship; then the length of the rod will not change. How do we know that perpendicular lengths do not change? The men can agree to make marks on each other's y-meter stick as they pass each other. By symmetry, the two marks must come at the same y- and y'-coordinates, since otherwise, when they get together to compare results, one mark will be above or below the other, and so we could tell who was really moving.

Now let us see what happens to the moving clock. Before the man took it aboard, he agreed that it was a nice, standard clock, and when he goes along in the space ship he will not see anything peculiar. If he did, he would know he was moving—if anything at all changed because of the motion, he could tell he was moving. But the principle of relativity says this is impossible in a uniformly moving system, so nothing has changed. On the other hand, when the external observer looks at the clock going by, he sees that the light, in going from mirror to mirror, is "really" taking a zigzag path, since the rod is moving sideways all the while. We have already analyzed such a zigzag motion in connection with the Michelson-Morley experiment. If in a given time the rod moves forward a distance proportional to u in Figure 3-3, the distance the light

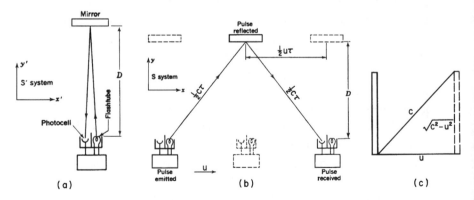

Figure 3-3. (a) A "light clock" at rest in the S' system. (b) The same clock, moving through the S system. (c) Illustration of the diagonal path taken by the light beam in a moving "light clock."

The Special Theory of Relativity

travels in the same time is proportional to c, and the vertical distance is therefore proportional to $\sqrt{c^2 - u^2}$.

That is, it takes a *longer time* for light to go from end to end in the moving clock than in the stationary clock. Therefore the apparent time between clicks is longer for the moving clock, in the same proportion as shown in the hypotenuse of the triangle (that is the source of the square root expressions in our equations). From the figure it is also apparent that the greater u is, the more slowly the moving clock appears to run. Not only does this particular kind of clock run more slowly, but if the theory of relativity is correct, any other clock, operating on any principle whatsoever, would also appear to run slower, and in the same proportion—we can say this without further analysis. Why is this so?

To answer the above question, suppose we had two other clocks made exactly alike with wheels and gears, or perhaps based on radioactive decay, or something else. Then we adjust these clocks so they both run in precise synchronism with our first clocks. When light goes up and back in the first clocks and announces its arrival with a click, the new models also complete some sort of cycle, which they simultaneously announce by some doubly coincident flash, or bong, or other signal. One of these clocks is taken into the space ship, along with the first kind. Perhaps *this* clock will not run slower, but will continue to keep the same time as its stationary counterpart, and thus disagree with the other moving clock. Ah no, if that should happen, the man in the ship could use this mismatch between his two clocks to determine the speed of his ship, which we have been supposing is impossible. *We need not know anything about the machinery* of the new clock that might cause the effect— we simply know that whatever the reason, it will appear to run slow, just like the first one.

Now if *all* moving clocks run slower, if no way of measuring time gives anything but a slower rate, we shall just have to say, in a certain sense, that *time itself* appears to be slower in a space ship. All the phenomena there—the man's pulse rate, his thought processes, the time he takes to light a cigar, how long it takes to grow up and get old—all these things must be slowed down in the same proportion, because he cannot tell he is moving. The biologists and medical men sometimes say it is not quite certain that the time it takes for a cancer to develop will be longer in a space ship, but from the viewpoint of a modern physicist it is nearly certain; otherwise one could use the rate of cancer development to determine the speed of the ship!

A very interesting example of the slowing of time with motion is furnished by mu-mesons (muons), which are particles that disintegrate spontaneously after an average lifetime of 2.2×10^{-6} sec. They come to the earth in cosmic rays, and can also be produced artificially in the laboratory. Some of them disintegrate in midair, but the remainder disintegrate only after they encounter a piece of material and stop. It is clear that in its short lifetime a muon cannot travel, even at the speed of light, much more than 600 meters. But although the muons are created at the top of the atmosphere, some 10 kilometers up, yet they are actually found in a laboratory down here, in cosmic rays. How can that be? The answer is that different muons move at various speeds, some of which are very close to the speed of light. While from their own point of view they live only about 2 μsec, from our point of view they live considerably longer—enough longer that they may reach the earth. The factor by which the time is increased has already been given as $1/\sqrt{1 - u^2/c^2}$. The average life has been measured quite accurately for muons of different velocities, and the values agree closely with the formula.

We do not know why the meson disintegrates or what its machinery is, but we do know its behavior satisfies the principle of relativity. That is the utility of the principle of relativity—it permits us to make predictions, even about things that otherwise we do not know much about. For example, before we have any idea at all about what makes the meson disintegrate, we can still predict that when it is moving at nine-tenths of the speed of light, the apparent length of time that it lasts is $(2.2 \times 10^{-6})/\sqrt{1 - 9^2/10^2}$ sec; and our prediction works—that is the good thing about it.

The Special Theory of Relativity

3-5 *The Lorentz contraction*

Now let us return to the Lorentz transformation (3.3) and try to get a better understanding of the relationship between the (x, y, z, t) and the (x', y', z', t') coordinate systems, which we shall call the S and S' systems, or Joe and Moe systems, respectively. We have already noted that the first equation is based on the Lorentz suggestion of contraction along the x-direction; how can we prove that a contraction takes place? In the Michelson-Morley experiment, we now appreciate that the *transverse* arm BC cannot change length, by the principle of relativity; yet the null result of the experiment demands that the *times* must be equal. So, in order for the experiment to give a null result, the longitudinal arm BE must appear shorter, by the square root $\sqrt{1 - u^2/c^2}$. What does this contraction mean, in terms of measurements made by Joe and Moe? Suppose that Moe, moving with the S' system in the x-direction, is measuring the x'-coordinate of some point with a meter stick. He lays the stick down x' times, so he thinks the distance is x' meters. From the viewpoint of Joe in the S system, however, Moe is using a foreshortened ruler, so the "real" distance measured is $x'\sqrt{1 - u^2/c^2}$ meters. Then if the S' system has travelled a distance ut away from the S system, the S observer would say that the same point, measured in his coordinates, is at a distance $x = x' \sqrt{1 - u^2/c^2} + ut$, or

$$x' = \frac{x - ut}{\sqrt{1 - u^2/c^2}},$$

which is the first equation of the Lorentz transformation.

3-6 *Simultaneity*

In an analogous way, because of the difference in time scales, the denominator expression is introduced into the fourth equation of the Lorentz transformation. The most interesting term in that equation is the ux/c^2 in the numerator, because that is quite new and unexpected. Now what does that mean? If we look at the situation carefully we see that events that occur at two separated places at the same time, as seen by Moe in S', do *not* happen at the same time as viewed by Joe in S. If one event occurs at point x_1 at time t_0 and the other event at x_2 and t_0 (the same time), we find that the two corresponding times t'_1 and t'_2 differ by an amount

$$t'_2 - t'_1 = \frac{u(x_1 - x_2)/c^2}{\sqrt{1 - u^2/c^2}} \, .$$

This circumstance is called "failure of simultaneity at a distance," and to make the idea a little clearer let us consider the following experiment.

Suppose that a man moving in a space ship (system S') has placed a clock at each end of the ship and is interested in making sure that the two clocks are in synchronism. How can the clocks be synchronized? There are many ways. One way, involving very little calculation, would be first to locate exactly the midpoint between the clocks. Then from this station we send out a light signal which will go both ways at the same speed and will arrive at both clocks, clearly, at the same time. This simultaneous arrival of the signals can be used to synchronize the clocks. Let us then suppose that the man in S' synchronizes his clocks by this particular method. Let us see whether an observer in system S would agree that the two clocks are synchronous. The man in S' has a right to believe they are, because he does not know that he is moving. But the man in S reasons that since the ship is moving forward, the clock in the front end was running away from the light signal, hence the light had to go more than halfway in order to catch up; the rear clock, however, was advancing to meet the light signal, so this distance was shorter. Therefore the signal reached the rear clock first, although the man in S' thought that the signals arrived simultaneously. We thus see that when a man in a space ship thinks the times at two locations are simultaneous, equal values of t' in his coordinate system must correspond to *different* values of t in the other coordinate system!

The Special Theory of Relativity

3-7 Four-vectors

Let us see what else we can discover in the Lorentz transformation. It is interesting to note that the transformation between the x's and t's is analogous in form to the transformation of the x's and y's that we studied in Chapter 1 for a rotation of coordinates. We then had

$$x' = x \cos \theta + y \sin \theta,$$
$$y' = y \cos \theta - x \sin \theta,$$

(3.8)

in which the new x' mixes the old x and y, and the new y' also mixes the old x and y; similarly, in the Lorentz transformation we find a new x' which is a mixture of x and t, and a new t' which is a mixture of t and x. So the Lorentz transformation is analogous to a rotation, only it is a "rotation" in *space and time*, which appears to be a strange concept. A check of the analogy to rotation can be made by calculating the quantity

$$x'^2 + y'^2 + z'^2 - c^2t'^2 = x^2 + y^2 + z^2 - c^2t^2.$$

(3.9)

In this equation the first three terms on each side represent, in three-dimensional geometry, the square of the distance between a point and the origin (surface of a sphere) which remains unchanged (invariant) regardless of rotation of the coordinate axes. Similarly, Eq. (3.9) shows that there is a certain combination which includes time, that is invariant to a Lorentz transformation. Thus, the analogy to a rotation is complete, and is of such a kind that vectors, i.e., quantities involving "components" which transform the same way as the coordinates and time, are also useful in connection with relativity.

Thus we contemplate an extension of the idea of vectors, which we have so far considered to have only space components, to include a time component. That is, we expect that there will be vectors with four components, three of which are like the components of an ordinary vector, and with these will be associated a fourth component, which is the analog of the time part.

This concept will be analyzed further in the next chapters, where we shall find that if the ideas of the preceding paragraph are applied to momentum, the transformation gives three space parts that are like ordinary momentum components, and a fourth component, the time part, which is the *energy*.

3-8 Relativistic dynamics

We are now ready to investigate, more generally, what form the laws of mechanics take under the Lorentz transformation. [We have thus far explained how length and time change, but not how we get the modified formula for m (Eq. 3.1). We shall do this in the next chapter.] To see the consequences of Einstein's modification of m for Newtonian mechanics, we start with the Newtonian law that force is the rate of change of momentum, or

$$\mathbf{F} = d(m\mathbf{v})/dt.$$

Momentum is still given by $m v$, but when we use the new m this becomes

$$\mathbf{p} = m\mathbf{v} = \frac{m_0\mathbf{v}}{\sqrt{1 - v^2/c^2}}. \tag{3.10}$$

This is Einstein's modification of Newton's laws. Under this modification, if action and reaction are still equal (which they may not be in detail, but are in the long run), there will be conservation of momentum in the same way as before, but the quantity that is being conserved is not the old $m v$ with its constant mass, but instead is the quantity shown in (3.10), which has the modified mass. When this change is made in the formula for momentum, conservation of momentum still works.

Now let us see how momentum varies with speed. In Newtonian mechanics it is proportional to the speed and, according to (3.10),

The Special Theory of Relativity

over a considerable range of speed, but small compared with c, it is nearly the same in relativistic mechanics, because the square-root expression differs only slightly from 1. But when v is almost equal to c, the square-root expression approaches zero, and the momentum therefore goes toward infinity.

What happens if a constant force acts on a body for a long time? In Newtonian mechanics the body keeps picking up speed until it goes faster than light. But this is impossible in relativistic mechanics. In relativity, the body keeps picking up, not speed, but momentum, which can continually increase because the mass is increasing. After a while there is practically no acceleration in the sense of a change of velocity, but the momentum continues to increase. Of course, whenever a force produces very little change in the velocity of a body, we say that the body has a great deal of inertia, and that is exactly what our formula for relativistic mass says (see Eq. 3.10)—it says that the inertia is very great when v is nearly as great as c. As an example of this effect, to deflect the high-speed electrons in the synchrotron that is used here at Caltech, we need a magnetic field that is 2000 times stronger than would be expected on the basis of Newton's laws. In other words, the mass of the electrons in the synchrotron is 2000 times as great as their normal mass, and is as great as that of a proton! That m should be 2000 times m_0 means that $1 - v^2/c^2$ must be 1/4,000,000, and that means that v^2/c^2 differs from 1 by one part in 4,000,000, or that v differs from c by one part in 8,000,000, so the electrons are getting pretty close to the speed of light. If the electrons and light were both to start from the synchrotron (estimated as 700 feet away) and rush out to Bridge Lab, which would arrive first? The light, of course, because light always travels faster.* How much earlier? That is too hard to tell—instead, we tell by what distance the light is ahead: it is about 1/1000 of an inch, or ¼ the thickness of a piece of paper!

---

* The electrons would actually win the race versus *visible* light because of the index of refraction of air. A gamma ray would make out better.

When the electrons are going that fast their masses are enormous, but their speed cannot exceed the speed of light.

Now let us look at some further consequences of relativistic change of mass. Consider the motion of the molecules in a small tank of gas. When the gas is heated, the speed of the molecules is increased, and therefore the mass is also increased and the gas is heavier. An approximate formula to express the increase of mass, for the case when the velocity is small, can be found by expanding $m_0/\sqrt{1 - v^2/c^2} = m_0(1 - v^2/c^2)^{-1/2}$ in a power series, using the binomial theorem. We get

$$m_0(1 - v^2/c^2)^{-1/2} = m_0(1 + \tfrac{1}{2}v^2/c^2 + \tfrac{3}{8}v^4/c^4 + \cdots).$$

We see clearly from the formula that the series converges rapidly when v is small, and the terms after the first two or three are negligible. So we can write

$$m \cong m_0 + \tfrac{1}{2}m_0 v^2\left(\frac{1}{c^2}\right) \tag{3.11}$$

in which the second term on the right expresses the increase of mass due to molecular velocity. When the temperature increases the v^2 increases proportionately, so we can say that the increase in mass is proportional to the increase in temperature. But since $\tfrac{1}{2}\,m_0 v^2$ is the kinetic energy in the old-fashioned Newtonian sense, we can also say that the increase in mass of all this body of gas is equal to the increase in kinetic energy divided by c^2, or $\Delta m = \Delta(\text{K.E.})/c^2$.

3-9 Equivalence of mass and energy

The above observation led Einstein to the suggestion that the mass of a body can be expressed more simply than by the formula (3.1), if we say that the mass is equal to the total energy content divided by c^2. If Eq. (3.11) is multiplied by c^2 the result is

$$mc^2 = m_0 c^2 + \tfrac{1}{2}m_0 v^2 + \cdots \tag{3.12}$$

Here, the term on the left expresses the total energy of a body, and we recognize the last term as the ordinary kinetic energy. Einstein

The Special Theory of Relativity

interpreted the large constant term, m_0c^2, to be part of the total energy of the body, an intrinsic energy known as the "rest energy."

Let us follow out the consequences of assuming, with Einstein, that *the energy of a body always equals mc^2*. As an interesting result, we shall find the formula (3.1) for the variation of mass with speed, which we have merely assumed up to now. We start with the body at rest, when its energy is m_0c^2. Then we apply a force to the body, which starts it moving and gives it kinetic energy; therefore, since the energy has increased, the mass has increased—this is implicit in the original assumption. So long as the force continues, the energy and the mass both continue to increase. We have already seen (Chapter 13*) that the rate of change of energy with time equals the force times the velocity, or

$$\frac{dE}{dt} = \mathbf{F} \cdot \mathbf{v}. \tag{3.13}$$

We also have (Chapter 9*, Eq. 9.1) that $F = d(mv)/dt$. When these relations are put together with the definition of E, Eq. (3.13) becomes

$$\frac{d(mc^2)}{dt} = \mathbf{v} \cdot \frac{d(m\mathbf{v})}{dt}. \tag{3.14}$$

We wish to solve this equation for m. To do this we first use the mathematical trick of multiplying both sides by $2m$, which changes the equation to

$$c^2(2m)\frac{dm}{dt} = 2mv\frac{d(mv)}{dt}. \tag{3.15}$$

We need to get rid of the derivatives, which can be accomplished by integrating both sides. The quantity $(2m)\,dm/dt$ can be recognized

---

* of the original *Lectures on Physics*, vol. I.

as the time derivative of m^2, and $(2m\mathbf{v}) \cdot d(m\mathbf{v})/dt$ is the time derivative of $(mv)^2$. So, Eq. (3.15) is the same as

$$c^2 \frac{d(m^2)}{dt} = \frac{d(m^2 v^2)}{dt}. \qquad (3.16)$$

If the derivatives of two quantities are equal, the quantities themselves differ at most by a constant, say C. This permits us to write

$$m^2 c^2 = m^2 v^2 + C. \qquad (3.17)$$

We need to define the constant C more explicitly. Since Eq. (3.17) must be true for all velocities, we can choose a special case where $v = 0$, and say that in this case the mass is m_0. Substituting these values into Eq. (3.17) gives

$$m_0^2 c^2 = 0 + C.$$

We can now use this value of C in Eq. (3.17), which becomes

$$m^2 c^2 = m^2 v^2 + m_0^2 c^2. \qquad (3.18)$$

Dividing by c^2 and rearranging terms gives

$$m^2(1 - v^2/c^2) = m_0^2,$$

from which we get

$$m = m_0/\sqrt{1 - v^2/c^2}. \qquad (3.19)$$

This is the formula (3.1), and is exactly what is necessary for the agreement between mass and energy in Eq. (3.12).

Ordinarily these energy changes represent extremely slight changes in mass, because most of the time we cannot generate much energy from a given amount of material; but in an atomic bomb of explosive energy equivalent to 20 kilotons of TNT, for example, it can be shown that the dirt after the explosion is lighter by 1 gram than the initial mass of the reacting material, because of the energy that was released, i.e., the released energy had a mass of 1 gram, according to the relationship $\Delta E = \Delta(mc^2)$. This theory of equivalence of mass and energy has been beautifully verified by experi-

The Special Theory of Relativity

ments in which matter is annihilated—converted totally to energy: An electron and a positron come together at rest, each with a rest mass m_0. When they come together they disintegrate and two gamma rays emerge, each with the measured energy of $m_0 c^2$. This experiment furnishes a direct determination of the energy associated with the existence of the rest mass of a particle.

Four

RELATIVISTIC ENERGY
AND MOMENTUM

4-1 Relativity and the philosophers

◈ In this chapter we shall continue to discuss the principle of
relativity of Einstein and Poincaré, as it affects our ideas of
physics and other branches of human thought.

Poincaré made the following statement of the principle of relativity: "According to the principle of relativity, the laws of physical
phenomena must be the same for a fixed observer as for an observer
who has a uniform motion of translation relative to him, so that we
have not, nor can we possibly have, any means of discerning
whether or not we are carried along in such a motion."

When this idea descended upon the world, it caused a great stir
among philosophers, particularly the "cocktail-party philosophers," who say, "Oh, it is very simple: Einstein's theory says all is
relative!" In fact, a surprisingly large number of philosophers, not
only those found at cocktail parties (but rather than embarrass
them, we shall just call them "cocktail-party philosophers"), will say,
"That all is relative is a consequence of Einstein, and it has profound influences on our ideas." In addition, they say "It has been
demonstrated in physics that phenomena depend upon your frame
of reference." We hear that a great deal, but it is difficult to find out

what it means. Probably the frames of reference that were originally referred to were the coordinate systems which we use in the analysis of the theory of relativity. So the fact that "things depend upon your frame of reference" is supposed to have had a profound effect on modern thought. One might well wonder why, because, after all, that things depend upon one's point of view is so simple an idea that it certainly cannot have been necessary to go to all the trouble of the physical relativity theory in order to discover it. That what one sees depends upon his frame of reference is certainly known to anybody who walks around, because he sees an approaching pedestrian first from the front and then from the back; there is nothing deeper in most of the philosophy which is said to have come from the theory of relativity than the remark that "A person looks different from the front than from the back." The old story about the elephant that several blind men describe in different ways is another example, perhaps, of the theory of relativity from the philosopher's point of view.

But certainly there must be deeper things in the theory of relativity than just this simple remark that "A person looks different from the front than from the back." Of course relativity is deeper than this, because *we can make definite predictions with it.* It certainly would be rather remarkable if we could predict the behavior of nature from such a simple observation alone.

There is another school of philosophers who feel very uncomfortable about the theory of relativity, which asserts that we cannot determine our absolute velocity without looking at something outside, and who would say, "It is obvious that one cannot measure his velocity without looking outside. It is self-evident that it is *meaningless* to talk about the velocity of a thing without looking outside; the physicists are rather stupid for having thought otherwise, but it has just dawned on them that this is the case. If only we philosophers had realized what the problems were that the physicists had, we could have decided immediately by brainwork that it is impossible to tell how fast one is moving without looking outside, and we could have made an enormous contribution to physics." These philosophers are always with us, struggling in the periphery to try to tell us

Relativistic Energy and Momentum

something, but they never really understand the subtleties and depths of the problem.

Our inability to detect absolute motion is a result of *experiment* and not a result of plain thought, as we can easily illustrate. In the first place, Newton believed that it was true that one could not tell how fast he is going if he is moving with uniform velocity in a straight line. In fact, Newton first stated the principle of relativity, and one quotation made in the last chapter was a statement of Newton's. Why then did the philosophers not make all this fuss about "all is relative," or whatever, in Newton's time? Because it was not until Maxwell's theory of electrodynamics was developed that there were physical laws that suggested that one *could* measure his velocity without looking outside; soon it was found *experimentally* that one could *not*.

Now, *is* it absolutely, definitely, philosophically *necessary* that one should not be able to tell how fast he is moving without looking outside? One of the consequences of relativity was the development of a philosophy which said, "You can only define what you can measure! Since it is self-evident that one cannot measure a velocity without seeing what he is measuring it relative to, therefore it is clear that there is no *meaning* to absolute velocity. The physicists should have realized that they can talk only about what they can measure." But *that is the whole problem:* whether or not one *can define* absolute velocity is the same as the problem of whether or not one *can detect in an experiment*, without looking outside, whether he is moving. In other words, whether or not a thing is measurable is not something to be decided *a priori* by thought alone, but something that can be decided only by experiment. Given the fact that the velocity of light is 186,000 mi/sec, one will find few philosophers who will calmly state that it is self-evident that if light goes 186,000 mi/sec inside a car, and the car is going 100,000 mi/sec, that the light also goes 186,000 mi/sec past an observer on the ground. That is a shocking fact to them; the very ones who claim it is obvious find, when you give them a specific fact, that it is not obvious.

Finally, there is even a philosophy which says that one cannot detect *any* motion except by looking outside. It is simply not true in physics. True, one cannot perceive a *uniform* motion in a *straight line*, but if the whole room were *rotating* we would certainly know it, for everybody would be thrown to the wall—there would be all kinds of "centrifugal" effects. That the earth is turning on its axis can be determined without looking at the stars, by means of the so-called Foucault pendulum, for example. Therefore it is not true that "all is relative"; it is only *uniform velocity* that cannot be detected without looking outside. Uniform *rotation* about a fixed axis *can* be. When this is told to a philosopher, he is very upset that he did not really understand it, because to him it seems impossible that one should be able to determine rotation about an axis without looking outside. If the philosopher is good enough, after some time he may come back and say, "I understand. We really do not have such a thing as absolute rotation; we are really rotating *relative to the stars*, you see. And so some influence exerted by the stars on the object must cause the centrifugal force."

Now, for all we know, that is true; we have no way, at the present time, of telling whether there would have been centrifugal force if there were no stars and nebulae around. We have not been able to do the experiment of removing all the nebulae and then measuring our rotation, so we simply do not know. We must admit that the philosopher may be right. He comes back, therefore, in delight and says, "It is absolutely necessary that the world ultimately turn out to be this way: *absolute* rotation means nothing; it is only *relative* to the nebulae." Then we say to him, "*Now*, my friend, is it or is it not obvious that uniform velocity in a straight line, *relative to the nebulae*, should produce no effects inside a car?" Now that the motion is no longer absolute, but is a motion *relative to the nebulae*, it becomes a mysterious question, and a question that can be answered only by experiment.

What, then, *are* the philosophic influences of the theory of relativity? If we limit ourselves to influences in the sense of *what kind of new ideas and suggestions* are made to the physicist by the principle of relativity, we could describe some of them as follows. The first discovery is, essentially, that even those ideas which have been held for a very long time and which have been very accurately verified might be wrong. It was a shocking discovery, of course, that Newton's laws are wrong, after all the years in which they seemed to be accurate. Of course it is clear, not that the experiments were wrong, but that they were done over only a limited range of velocities, so small that the relativistic effects would not have been evident. But nevertheless, we now have a much more humble point of view of our physical laws—everything *can* be wrong!

Secondly, if we have a set of "strange" ideas, such as that time goes slower when one moves, and so forth, whether we *like* them or do *not* like them is an irrelevant question. The only relevant question is whether the ideas are consistent with what is found experimentally. In other words, the "strange ideas" need only agree with *experiment*, and the only reason that we have to discuss the behavior of clocks and so forth is to demonstrate that although the notion of the time dilation is strange, it is *consistent* with the way we measure time.

Finally, there is a third suggestion which is a little more technical but which has turned out to be of enormous utility in our study of other physical laws, and that is to *look at the symmetry of the laws* or, more specifically, to look for the ways in which the laws can be transformed and leave their form the same. When we discussed the theory of vectors, we noted that the fundamental laws of motion are not changed when we rotate the coordinate system, and now we learn that they are not changed when we change the space and time variables in a particular way, given by the Lorentz transformation. So this idea of studying the patterns or operations under which the fundamental laws are not changed has proved to be a very useful one.

4-2 *The twin paradox*

To continue our discussion of the Lorentz transformation and relativistic effects, we consider a famous so-called "paradox" of Peter and Paul, who are supposed to be twins, born at the same time. When they are old enough to drive a space ship, Paul flies away at very high speed. Because Peter, who is left on the ground, sees Paul going so fast, all of Paul's clocks appear to go slower, his heart beats go slower, his thoughts go slower, everything goes slower, from Peter's point of view. Of course, Paul notices nothing unusual, but if he travels around and about for a while and then comes back, he will be younger than Peter, the man on the ground! That is actually right; it is one of the consequences of the theory of relativity which has been clearly demonstrated. Just as the mu-mesons last longer when they are moving, so also will Paul last longer when he is moving. This is called a "paradox" only by the people who believe that the principle of relativity means that *all motion* is relative; they say, "Heh, heh, heh, from the point of view of Paul, can't we say that *Peter* was moving and should therefore appear to age more slowly? By symmetry, the only possible result is that both should be the same age when they meet." But in order for them to come back together and make the comparison, Paul must either stop at the end of the trip and make a comparison of clocks or, more simply, he has to come back, and the one who comes back must be the man who was moving, and he knows this, because he had to turn around. When he turned around, all kinds of unusual things happened in his space ship—the rockets went off, things jammed up against one wall, and so on—while Peter felt nothing.

So the way to state the rule is to say that *the man who has felt the accelerations*, who has seen things fall against the walls, and so on, is the one who would be the younger; that is the difference between them in an "absolute" sense, and it is certainly correct. When we discussed the fact that moving mu-mesons live longer, we used as an example their straight-line motion in the atmosphere. But we can also make mu-mesons in a laboratory and cause them to go in a curve with a magnet, and even under this accelerated motion, they last exactly as much longer as they do when they are moving in a straight line. Although no one has arranged an experiment explicitly so that we can get rid of the paradox, one could compare a mu-meson which is left standing with one that had gone around a

Relativistic Energy and Momentum

complete circle, and it would surely be found that the one that went around the circle lasted longer. Although we have not actually carried out an experiment using a complete circle, it is really not necessary, of course, because everything fits together all right. This may not satisfy those who insist that every single fact be demonstrated directly, but we confidently predict the result of the experiment in which Paul goes in a complete circle.

4-3 *Transformation of velocities*

The main difference between the relativity of Einstein and the relativity of Newton is that the laws of transformation connecting the coordinates and times between relatively moving systems are different. The correct transformation law, that of Lorentz, is

$$x' = \frac{x - ut}{\sqrt{1 - u^2/c^2}},$$
$$y' = y,$$
$$z' = z, \tag{4.1}$$
$$t' = \frac{t - ux/c^2}{\sqrt{1 - u^2/c^2}}.$$

These equations correspond to the relatively simple case in which the relative motion of the two observers is along their common x-axes. Of course other directions of motion are possible, but the most general Lorentz transformation is rather complicated, with all four quantities mixed up together. We shall continue to use this simpler form, since it contains all the essential features of relativity.

Let us now discuss more of the consequences of this transformation. First, it is interesting to solve these equations in reverse. That is, here is a set of linear equations, four equations with four unknowns, and they can be solved in reverse, for x, y, z, t in terms of x', y', z', t'. The result is very interesting, since it tells us how a system of coordinates "at rest" looks from the point of view of one that is "moving." Of course, since the motions are relative and of uniform velocity, the man who is "moving" can say, if he wishes, that it is really the other fellow who is moving and he himself who is at rest. And since he is moving in the opposite direction, he should get the same transformation, but with the opposite sign of velocity. That is precisely what we find by manipulation, so that is consistent. If it did not come out that way, we would have real cause to worry!

$$x = \frac{x' + ut'}{\sqrt{1 - u^2/c^2}},$$
$$y = y',$$
$$z = z',$$
$$t = \frac{t' + ux'/c^2}{\sqrt{1 - u^2/c^2}}.$$

(4.2)

Next we discuss the interesting problem of the addition of velocities in relativity. We recall that one of the original puzzles was that light travels at 186,000 mi/sec in all systems, even when they are in relative motion. This is a special case of the more general problem exemplified by the following. Suppose that an object inside a space ship is going at 100,000 mi/sec and the space ship itself is going at 100,000 mi/sec; how fast is the object inside the space ship moving from the point of view of an observer outside? We might want to say 200,000 mi/sec, which is faster than the speed of light. This is very unnerving, because it is not supposed to be going faster than the speed of light! The general problem is as follows.

Let us suppose that the object inside the ship, from the point of view of the man inside, is moving with velocity v, and that the space ship itself has a velocity u with respect to the ground. We want to know with what velocity v_x this object is moving from the point of view of the man on the ground. This is, of course, still but a special case in which the motion is in the x-direction. There will also be a transformation for velocities in the y-direction, or for any angle; these can be worked out as needed. Inside the space ship the velocity is $v_{x'}$, which means that the displacement x is equal to the velocity times the time:

$$x' = v_{x'}t'.$$

(4.3)

Relativistic Energy and Momentum

Now we have only to calculate what the position and time are from the point of view of the outside observer for an object which has the relation (4.2) between x' and t'. So we simply substitute (4.3) into (4.2), and obtain

$$x = \frac{v_{x'}t' + ut'}{\sqrt{1 - u^2/c^2}} . \tag{4.4}$$

But here we find x expressed in terms of t'. In order to get the velocity as seen by the man on the outside, we must divide *his distance* by *his time*, not by the *other man's time*! So we must also calculate the *time* as seen from the outside, which is

$$t = \frac{t' + u(v_{x'}t')/c^2}{\sqrt{1 - u^2/c^2}} . \tag{4.5}$$

Now we must find the ratio of x to t, which is

$$v_x = \frac{x}{t} = \frac{u + v_{x'}}{1 + uv_{x'}/c^2} , \tag{4.6}$$

the square roots having cancelled. This is the law that we seek: the resultant velocity, the "summing" of two velocities, is not just the algebraic sum of two velocities (we know that it cannot be or we get in trouble), but is "corrected" by $1 + uv/c^2$.

Now let us see what happens. Suppose that you are moving inside the space ship at half the speed of light, and that the space ship itself is going at half the speed of light. Thus u is $\frac{1}{2}c$ and v is $\frac{1}{2}c$, but in the denominator uv is one-fourth, so that

$$v = \frac{\frac{1}{2}c + \frac{1}{2}c}{1 + \frac{1}{4}} = \frac{4c}{5} .$$

So, in relativity, "half" and "half" does not make "one," it makes only "4/5." Of course low velocities can be added quite easily in the familiar way, because so long as the velocities are small compared with the speed of light we can forget about the $(1 + uv/c^2)$ factor; but things are quite different and quite interesting at high velocity.

Let us take a limiting case. Just for fun, suppose that inside the space ship the man was observing *light itself*. In other words, $v = c$, and yet the space ship is moving. How will it look to the man on the ground? The answer will be

$$v = \frac{u + c}{1 + uc/c^2} = c\frac{u + c}{u + c} = c.$$

Therefore, if something is moving at the speed of light inside the ship, it will appear to be moving at the speed of light from the point of view of the man on the ground too! This is good, for it is, in fact, what the Einstein theory of relativity was designed to do in the first place—so it had *better* work!

Of course, there are cases in which the motion is not in the direction of the uniform translation. For example, there may be an object inside the ship which is just moving "upward" with the velocity $v_{y'}$ with respect to the ship, and the ship is moving "horizontally." Now, we simply go through the same thing, only using y's instead of x's, with the result

$$y = y' = v_{y'}t',$$

so that if $v_{x'} = 0$,

$$v_y = \frac{y}{t} = v_{y'}\sqrt{1 - u^2/c^2}. \tag{4.7}$$

Thus a sidewise velocity is no longer $v_{y'}$, but $v_{y'}\sqrt{1 - u^2/c^2}$. We found this result by substituting and combining the transformation equations, but we can also see the result directly from the principle of relativity for the following reason (it is always good to look again to see whether we can see the reason). We have already (Figure 3-3) discussed how a possible clock might work when it is moving; the light appears to travel at an angle at the speed c in the fixed system, while it simply goes vertically with the same speed in the moving system. We found that the *vertical component* of the velocity in the fixed system is less than that of light by the factor $\sqrt{1 - u^2/c^2}$ (see Eq. 3-3). But now suppose that we let a material particle go back and forth in this same "clock," but at some integral fraction $1/n$ of the speed of light (Figure 4-1). Then when the particle has gone back and forth once, the light will have gone exactly n times. That is, each "click" of the "particle" clock will coincide with each nth

Relativistic Energy and Momentum

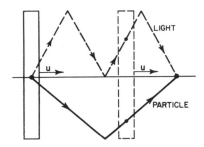

Figure 4-1. Trajectories described by a light ray and particle inside a moving clock.

"click" of the light clock. *This fact must still be true when the whole system is moving,* because the physical phenomenon of coincidence will be a coincidence in any frame. Therefore, since the speed c_y is less than the speed of light, the speed v_y of the particle must be slower than the corresponding speed by the same square-root ratio! That is why the square root appears in any vertical velocity.

4-4 Relativistic mass

We learned in the last chapter that the mass of an object increases with velocity, but no demonstration of this was given, in the sense that we made no arguments analogous to those about the way clocks have to behave. However, we *can* show that, as a consequence of relativity plus a few other reasonable assumptions, the mass must vary in this way. (We have to say "a few other assumptions" because we cannot prove anything unless we have some laws which we assume to be true, if we expect to make meaningful deductions.) To avoid the need to study the transformation laws of force, we shall analyze a *collision*, where we need know nothing about the laws of force, except that we shall assume the conservation of momentum and energy. Also, we shall assume that the momentum of a particle which is moving is a vector and is always directed in the direction of the velocity. However, we shall not assume that the momentum is a *constant* times the velocity, as Newton did, but only that it is some *function* of velocity. We thus write the momentum vector as a certain coefficient times the vector velocity:

$$\mathbf{p} = m_v\mathbf{v}. \qquad (4.8)$$

We put a subscript v on the coefficient to remind us that it is a function of velocity, and we shall agree to call this coefficient m_v the "mass." Of course, when the velocity is small, it is the same mass that we would measure in the slow-moving experiments that we are used to. Now we shall try to demonstrate that the formula for m_v must be $m_0/\sqrt{1 - v^2/c^2}$, by arguing from the principle of relativity that the laws of physics must be the same in every coordinate system.

Suppose that we have two particles, like two protons, that are absolutely equal, and they are moving toward each other with exactly equal velocities. Their total momentum is zero. Now what can happen? After the collision, their directions of motion must be exactly opposite to each other, because if they are not exactly opposite, there will be a nonzero total vector momentum, and momentum would not have been conserved. Also they must have the same speeds, since they are exactly similar objects; in fact, they must have the same speed they started with, since we suppose that the energy is conserved in these collisions. So the diagram of an elastic collision, a reversible collision, will look like Figure 4-2(a): all the arrows are the same length, all the speeds are equal. We shall suppose that such collisions can always be arranged, that any angle θ can occur, and that any speed could be used in such a collision. Next, we notice that this same collision can be viewed differently by turning the axes, and just for convenience we *shall* turn the axes, so that the horizontal splits it evenly, as in Figure 4-2(b). It is the same collision redrawn, only with the axes turned.

Now here is the real trick: let us look at this collision from the point of view of someone riding along in a car that is moving with a speed equal to the horizontal component of the velocity of one particle.

Relativistic Energy and Momentum

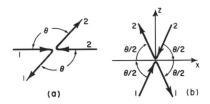

Figure 4-2. Two views of an elastic collision between equal objects moving at the same speed in opposite directions.

Then how does the collision look? It looks as though particle 1 is just going straight up, because it has lost its horizontal component, and it comes straight down again, also because it does not have that component. That is, the collision appears as shown in Fig. 4-3(a). Particle 2, however, was going the other way, and as we ride past it appears to fly by at some terrific speed and at a smaller angle, but we can appreciate that the angles before and after the collision are the *same*. Let us denote by u the horizontal component of the velocity of particle 2, and by w the vertical velocity of particle 1.

Now the question is, what is the vertical velocity $u \tan \alpha$? If we knew that, we could get the correct expression for the momentum, using the law of conservation of momentum in the vertical direction. Clearly, the horizontal component of the momentum is conserved: it is the same before and after the collision for both particles, and is zero for particle 1. So we need use the conservation law only for the upward velocity $u \tan \alpha$. But we *can* get the upward velocity, simply by looking at the same collision going the other way! If we look at the collision of Figure 4-3(a) from a car to the left moving with speed u, we see the same collision, except "turned over," as shown in Figure 4-3(b). Now particle 2 is the one that goes up and down with speed w, and particle 1 has picked up the horizontal speed u. Of course, now we *know* what the velocity $u \tan \alpha$ is: it is $w\sqrt{1 - u^2/c^2}$ (see Eq. 4.7). We know that the change in the vertical momentum of the vertically moving particle is

$$\Delta p = 2m_w w$$

SIX NOT-SO-EASY PIECES

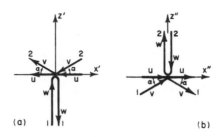

Figure 4-3. Two more views of the collision, from moving cars.

(2, because it moves up and back down). The obliquely moving particle has a certain velocity v whose components we have found to be u and $w\sqrt{1 - u^2/c^2}$, and whose mass is m_v. The change in *vertical* momentum of this particle is therefore $\Delta p' = 2m_v w \sqrt{1 - u^2/c^2}$ because, in accordance with our assumed law (4.8), the momentum component is always the mass corresponding to the magnitude of the velocity times the component of the velocity in the direction of interest. Thus in order for the total momentum to be zero the vertical momenta must cancel and the ratio of the mass moving with speed v and the mass moving with speed w must therefore be

$$\frac{m_w}{m_v} = \sqrt{1 - u^2/c^2}. \qquad (4.9)$$

Let us take the limiting case that w is infinitesimal. If w is very tiny indeed, it is clear that v and u are practically equal. In this case, $m_w \to m_0$ and $m_v \to m_u$. The grand result is

$$m_u = \frac{m_0}{\sqrt{1 - u^2/c^2}}. \qquad (4.10)$$

It is an interesting exercise now to check whether or not Eq. (4.9) is indeed true for arbitrary values of w, assuming that Eq. (4.10) is the right formula for the mass. Note that the velocity v needed in Eq. (4.9) can be calculated from the right-angle triangle:

$$v^2 = u^2 + w^2(1 - u^2/c^2).$$

Relativistic Energy and Momentum

Figure 4-4. Two views of an inelastic collision between equally massive objects.

It will be found to check out automatically, although we used it only in the limit of small w.

Now, let us accept that momentum is conserved and that the mass depends upon the velocity according to (4.10) and go on to find what else we can conclude. Let us consider what is commonly called an *inelastic collision*. For simplicity, we shall suppose that two objects of the same kind, moving oppositely with equal speeds w, hit each other and stick together, to become some new, stationary object, as shown in Figure 4-4(a). The mass m of each corresponds to w, which, as we know, is $m_0/\sqrt{1 - w^2/c^2}$. If we assume the conservation of momentum and the principle of relativity, we can demonstrate an interesting fact about the mass of the new object which has been formed. We imagine an infinitesimal velocity u at right angles to w (we can do the same with finite values of u, but it is easier to understand with an infinitesimal velocity), then look at this same collision as we ride by in an elevator at the velocity $-u$. What we see is shown in Figure 4-4(b). The composite object has an unknown mass M. Now object 1 moves with an upward component of velocity u and a horizontal component which is practically equal to w, and so also does object 2. After the collision we have the mass M moving upward with velocity u, considered very small compared with the speed of light, and also small compared with w. Momentum must be conserved, so let us estimate the momentum in the upward direction before and after the collision. Before the collision we have $p \sim 2m_w u$, and after the collision the momentum is evidently $p' = M_u u$, but M_u is essentially the same as M_0 because u is so small. These momenta must be equal because of the conservation of momentum, and therefore

$$M_0 = 2m_w. \tag{4.11}$$

The mass of the object which is formed when two equal objects collide must be twice the mass of the objects which come together. You might say, "Yes, of course, that is the conservation of mass." But not "Yes, of course," so easily, because *these masses have been enhanced* over the masses that they would be if they were standing still, yet they still contribute, to the total M, not the mass they have when standing still, but *more*. Astonishing as that may seem, in order for the conservation of momentum to work when two objects come together, the mass that they form must be greater than the rest masses of the objects, even though the objects are at rest after the collision!

4-5 Relativistic energy

In the last chapter we demonstrated that as a result of the dependence of the mass on velocity and Newton's laws, the changes in the kinetic energy of an object resulting from the total work done by the forces on it always comes out to be

$$\Delta T = (m_u - m_0)c^2 = \frac{m_0 c^2}{\sqrt{1 - u^2/c^2}} - m_0 c^2. \tag{4.12}$$

We even went further, and guessed that the total energy is the total mass times c^2. Now we continue this discussion.

Suppose that our two equally massive objects that collide can still be "seen" inside M. For instance, a proton and a neutron are "stuck together," but are still moving about inside of M. Then, although we might at first expect the mass M to be $2m_0$, we have found that it is not $2m_0$, but $2m_w$. Since $2m_w$ is what is put in, but $2m_0$ are the rest masses of the things inside, the *excess* mass of the composite object is equal to the kinetic energy brought in. This means, of course, that *energy has inertia*. In the last chapter we discussed the heating of a gas, and showed that because the gas molecules are moving and moving things are heavier, when we put energy into the gas its

Relativistic Energy and Momentum

molecules move faster and so the gas gets heavier. But in fact the argument is completely general, and our discussion of the inelastic collision shows that the mass is there whether or not it is *kinetic* energy. In other words, if two particles come together and produce potential or any other form of energy; if the pieces are slowed down by climbing hills, doing work against internal forces, or whatever; then it is still true that the mass is the total energy that has been put in. So we see that the conservation of mass which we have deduced above is equivalent to the conservation of energy, and therefore there is no place in the theory of relativity for strictly inelastic collisions, as there was in Newtonian mechanics. According to Newtonian mechanics it is all right for two things to collide and so form an object of mass $2m_0$ which is in no way distinct from the one that would result from putting them together slowly. Of course we know from the law of conservation of energy that there is more kinetic energy inside, but that does not affect the mass, according to Newton's laws. But now we see that this is impossible; because of the kinetic energy involved in the collision, the resulting object will be *heavier*; therefore, it will be a *different* object. When we put the objects together gently they make something whose mass is $2m_0$; when we put them together forcefully, they make something whose mass is greater. When the mass is different, we can *tell* that it is different. So, necessarily, the conservation of energy must go along with the conservation of momentum in the theory of relativity.

This has interesting consequences. For example, suppose that we have an object whose mass M is measured, and suppose something happens so that it flies into two equal pieces moving with speed w, so that they each have a mass m_w. Now suppose that these pieces encounter enough material to slow them up until they stop; then they will have mass m_0. How much energy will they have given to the material when they have stopped? Each will give an amount $(m_w - m_0)c^2$, by the theorem that we proved before. This much energy is left in the material in some form, as heat, potential energy, or whatever. Now $2m_w = M$, so the liberated energy is $E = (M - 2m_0)c^2$. This equation was used to estimate how much energy would be liberated under fission in the atomic bomb, for example. (Although the fragments are not exactly equal, they are nearly equal.) The mass of the uranium atom was known—it had been measured ahead of time—and the atoms into which it split, iodine, xenon, and so on, all were of known mass. By masses, we do not mean the masses while the atoms are moving, we mean the masses when the atoms are *at rest*. In other words, both M and m_0 are known. So by subtracting the two numbers one can calculate how much energy will be released if M can be made to split in "half." For this reason poor old Einstein was called the "father" of the atomic bomb in all the newspapers. Of course, all that meant was that he could tell us ahead of time how much energy would be released if we told him what process would occur. The energy that should be liberated when an atom of uranium undergoes fission was estimated about six months before the first direct test, and as soon as the energy was in fact liberated, someone measured it directly (and if Einstein's formula had not worked, they would have measured it anyway), and the moment they measured it they no longer needed the formula. Of course, we should not belittle Einstein, but rather should criticize the newspapers and many popular descriptions of what causes what in the history of physics and technology. The problem of how to get the thing to occur in an effective and rapid manner is a completely different matter.

The result is just as significant in chemistry. For instance, if we were to weigh the carbon dioxide molecule and compare its mass with that of the carbon and the oxygen, we could find out how much energy would be liberated when carbon and oxygen form carbon dioxide. The only trouble here is that the differences in masses are so small that it is technically very difficult to do.

Now let us turn to the question of whether we should add $m_0 c^2$ to the kinetic energy and say from now on that the total energy of an object is mc^2. First, if we can still *see* the component pieces of rest mass m_0 inside M, then we could say that some of the mass M of the compound object is the mechanical rest mass of the parts, part of it is kinetic energy of the parts, and part of it is potential energy of the

Relativistic Energy and Momentum

parts. But we have discovered, in nature, particles of various kinds which undergo reactions just like the one we have treated above, in which with all the study in the world, *we cannot see the parts inside.* For instance, when a K-meson disintegrates into two pions it does so according to the law (4.11), but the idea that a K is made out of 2 π's is a useless idea, because it also disintegrates into 3 π's!

Therefore we have a *new idea*: we do not have to know what things are made of inside; we cannot and need not identify, inside a particle, which of the energy is rest energy of the parts into which it is going to disintegrate. It is not convenient and often not possible to separate the total mc^2 energy of an object into rest energy of the inside pieces, kinetic energy of the pieces, and potential energy of the pieces; instead, we simply speak of the *total energy* of the particle. We "shift the origin" of energy by adding a constant m_0c^2 to everything, and say that the total energy of a particle is the mass in motion times c^2, and when the object is standing still, the energy is the mass at rest times c^2.

Finally, we find that the velocity v, momentum P, and total energy E are related in a rather simple way. That the mass in motion at speed v is the mass m_0 at rest divided by $\sqrt{1 - v^2/c^2}$, surprisingly enough, is rarely used. Instead, the following relations are easily proved, and turn out to be very useful:

$$E^2 - P^2c^2 = m_0^2c^4 \tag{4.13}$$

and

$$Pc = Ev/c. \tag{4.14}$$

Five

SPACE-TIME

5-1 *The geometry of space-time*

The theory of relativity shows us that the relationships of positions and times as measured in one coordinate system and another are not what we would have expected on the basis of our intuitive ideas. It is very important that we thoroughly understand the relations of space and time implied by the Lorentz transformation, and therefore we shall consider this matter more deeply in this chapter.

The Lorentz transformation between the positions and times (x, y, z, t) as measured by an observer "standing still," and the corresponding coordinates and time (x', y', z', t') measured inside a "moving" space ship, moving with velocity u are

$$x' = \frac{x - ut}{\sqrt{1 - u^2/c^2}},$$

$$y' = y,$$

$$z' = z,$$

$$t' = \frac{t - ux/c^2}{\sqrt{1 - u^2/c^2}}.$$

(5.1)

Let us compare these equations with Eq. (1.5), which also relates measurements in two systems, one of which in this instance is *rotated* relative to the other:

$$x' = x \cos \theta + y \sin \theta,$$
$$y' = y \cos \theta - x \sin \theta, \qquad (5.2)$$
$$z' = z.$$

In this particular case, Moe and Joe are measuring with axes having an angle θ between the x'- and x-axes. In each case, we note that the "primed" quantities are "mixtures" of the "unprimed" ones: the new x' is a mixture of x and y, and the new y' is also a mixture of x and y.

An analogy is useful: When we look at an object, there is an obvious thing we might call the "apparent width," and another we might call the "depth." But the two ideas, width and depth, are not *fundamental* properties of the object, because if we step aside and look at the same thing from a different angle, we get a different width and a different depth, and we may develop some formulas for computing the new ones from the old ones and the angles involved. Equations (5.2) are these formulas. One might say that a given depth is a kind of "mixture" of all depth and all width. If it were impossible ever to move, and we always saw a given object from the same position, then this whole business would be irrelevant—we would always see the "true" width and the "true" depth, and they would appear to have quite different qualities, because one appears as a subtended optical angle and the other involves some focusing of the eyes or even intuition; they would seem to be very different things and would never get mixed up. It is because we *can* walk around that we realize that depth and width are, somehow or other, just two different aspects of the same thing.

Can we not look at the Lorentz transformations in the same way? Here also we have a mixture—of positions and the time. A difference between a space measurement and a time measurement produces a new space measurement. In other words, in the space measurements of one man there is mixed in a little bit of the time, as seen by the other. Our analogy permits us to generate this idea: The "reality" of an object that we are looking at is somehow greater (speaking crudely and intuitively) than its "width" and its "depth"

Space-Time

because *they* depend upon *how* we look at it; when we move to a new position, our brain immediately recalculates the width and the depth. But our brain does not immediately recalculate coordinates and time when we move at high speed, because we have had no effective experience of going nearly as fast as light to appreciate the fact that time and space are also of the same nature. It is as though we were always stuck in the position of having to look at just the width of something, not being able to move our heads appreciably one way or the other; if we could, we understand now, we would see some of the other man's time—we would see "behind," so to speak, a little bit.

Thus we shall try to think of objects in a new kind of world, of space and time mixed together, in the same sense that the objects in our ordinary space-world are real, and can be looked at from different directions. We shall then consider that objects occupying space and lasting for a certain length of time occupy a kind of a "blob" in a new kind of world, and that we look at this "blob" from different points of view when we are moving at different velocities. This new world, this geometrical entity in which the "blobs" exist by occupying position and taking up a certain amount of time, is called *space-time*. A given point (x, y, z, t) in space-time is called an *event*. Imagine, for example, that we plot the x-positions horizontally, y and z in two other directions, both mutually at "right angles" and at "right angles" to the paper (!), and time, vertically. Now, how does a moving particle, say, look on such a diagram? If the particle is standing still, then it has a certain x, and as time goes on, it has the same x, the same x, the same x; so its "path" is a line that runs parallel to the t-axis (Figure 5-1 (a)). On the other hand, if it drifts outward, then as the time goes on x increases (Figure 5-1 (b)). So a particle, for example, which starts to drift out and then slows up should have a motion something like that shown in Figure 5-1 (c). A particle, in other words, which is permanent and does not disintegrate is represented by a line in space-time. A particle which disintegrates would be represented by a forked line, because it would turn into two other things which would start from that point.

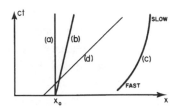

Figure 5-1. Three particle paths in space-time: (a) a particle at rest at $x = x_0$; (b) a particle which starts at $x = x_0$ and moves with constant speed; (c) a particle which starts at high speed but slows down.

What about light? Light travels at the speed c, and that would be represented by a line having a certain fixed slope (Figure 5-1 (d)).

Now according to our new idea, if a given event occurs to a particle, say if it suddenly disintegrates at a certain space-time point into two new ones which follow some new tracks, and this interesting event occurred at a certain value of x and a certain value of t, then we would expect that, if this makes any sense, we just have to take a new pair of axes and turn them, and that will give us the new t and the new x in our new system, as shown in Figure 5-2(a). But this is wrong, because Eq. (5.1) is not *exactly* the same mathematical transformation as Eq. (5.2). Note, for example, the difference in sign between the two, and the fact that one is written in terms of cos θ and sin θ, while the other is written with algebraic quantities. (Of course, it is not impossible that the algebraic quantities could be written as cosine and sine, but actually they cannot.) But still, the

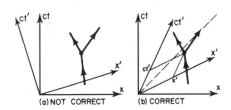

Figure 5-2. Two views of a disintegrating particle.

Space-Time

two expressions *are* very similar. As we shall see, it is not really possible to think of space-time as a real, ordinary geometry because of that difference in sign. In fact, although we shall not emphasize this point, it turns out that a man who is moving has to use a set of axes which are inclined equally to the light ray, using a special kind of projection parallel to the x'- and t'-axes, for his x' and t', as shown in Figure 5-2(b). We shall not deal with the geometry, since it does not help much; it is easier to work with the equations.

5-2 *Space-time intervals*

Although the geometry of space-time is not Euclidean in the ordinary sense, there *is* a geometry which is very similar, but peculiar in certain respects. If this idea of geometry is right, there ought to be some functions of coordinates and time which are independent of the coordinate system. For example, under ordinary rotations, if we take two points, one at the origin, for simplicity, and the other one somewhere else, both systems would have the same origin, and the distance from here to the other point is the same in both. That is one property that is independent of the particular way of measuring it. The square of the distance is $x^2 + y^2 + z^2$. Now what about space-time? It is not hard to demonstrate that we have here, also, something which stays the same, namely, the combination $c^2t^2 - x^2 - y^2 - z^2$ is the same before and after the transformation:

$$c^2t'^2 - x'^2 - y'^2 - z'^2 = c^2t^2 - x^2 - y^2 - z^2. \qquad (5.3)$$

This quantity is therefore something which, like the distance, is "real" in some sense; it is called the *interval* between the two space-time points, one of which is, in this case, at the origin. (Actually, of course, it is the interval squared, just as $x^2 + y^2 + z^2$ is the distance squared.) We give it a different name because it is in a different geometry, but the interesting thing is only that some signs are reversed and there is a c in it.

Let us get rid of the c; that is an absurdity if we are going to have a wonderful space with x's and y's that can be interchanged. One of the confusions that could be caused by someone with no experience would be to measure widths, say, by the angle subtended at the eye, and measure depth in a different way, like the strain on the muscles needed to focus them, so that the depths would be measured in feet and the widths in meters. Then one would get an enormously complicated mess of equations in making transformations such as (5.2), and would not be able to see the clarity and simplicity of the thing for a very simple technical reason, that the same thing is being measured in two different units. Now in Eqs. (5.1) and (5.3) nature is telling us that time and space are equivalent; time becomes space; *they should be measured in the same units*. What distance is a "second"? It is easy to figure out from (5.3) what it is. It is 3×10^8 meters, *the distance that light would go in one second*. In other words, if we were to measure all distances and times in the same units, seconds, then our unit of distance would be 3×10^8 meters, and the equations would be simpler. Or another way that we could make the units equal is to measure time in meters. What is a meter of time? A meter of time is the time it takes for light to go one meter, and is therefore $\frac{1}{3} \times 10^{-8}$ sec, or 3.3 billionths of a second! We would like, in other words, to put all our equations in a system of units in which $c = 1$. If time and space are measured in the same units, as suggested, then the equations are obviously much simplified. They are

$$x' = \frac{x - ut}{\sqrt{1 - u^2}},$$

$$y' = y,$$

$$z' = z, \qquad\qquad (5.4)$$

$$t' = \frac{t - ux}{\sqrt{1 - u^2}},$$

$$t'^2 - x'^2 - y'^2 - z'^2 = t^2 - x^2 - y^2 - z^2. \qquad (5.5)$$

If we are ever unsure or "frightened" that after we have this system with $c = 1$ we shall never be able to get our equations right again, the answer is quite the opposite. It is much easier to remember them without the c's in them, and it is always easy to put the c's back, by looking after the dimensions. For instance, in $\sqrt{1 - u^2}$, we know that we cannot subtract a velocity squared, which has units, from the pure number 1, so we know that we must divide u^2 by c^2 in order to make that unitless, and that is the way it goes.

The difference between space-time and ordinary space, and the character of an interval as related to the distance, is very interesting. According to formula (5.5), if we consider a point which in a given coordinate system had zero time, and only space, then the interval squared would be negative and we would have an imaginary interval, the square root of a negative number. Intervals can be either real or imaginary in the theory. The square of an interval may be either positive or negative, unlike distance, which has a positive square. When an interval is imaginary, we say that the two points have a *space-like interval* between them (instead of imaginary), because the interval is more like space than like time. On the other hand, if two objects are at the same place in a given coordinate system, but differ only in time, then the square of the time is positive and the distances are zero and the interval squared is positive; this is called a *time-like interval.* In our diagram of space-time, therefore, we would have a representation something like this: at 45° there are two lines (actually, in four dimensions these will be "cones," called light cones) and points on these lines are all at zero interval from the origin. Where light goes from a given point is always separated from it by a zero interval, as we see from Eq. (5.5). Incidentally, we have just proved that if light travels with speed c in one system, it travels with speed c in another, for if the interval is the same in both systems, i.e., zero in one and zero in the other, then to state that the propagation speed of light is invariant is the same as saying that the interval is zero.

5-3 *Past, present, and future*

The space-time region surrounding a given space-time point can be separated into three regions, as shown in Figure 5-3. In one region we have space-like intervals, and in two regions, time-like intervals.

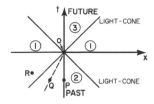

Figure 5-3. The space-time region surrounding a point at the origin.

Physically, these three regions into which space-time around a given point is divided have an interesting physical relationship to that point: a physical object or a signal can get from a point in region 2 to the event O by moving along at a speed less than the speed of light. Therefore events in this region can affect the point O, can have an influence on it from the past. In fact, of course, an object at P on the negative t-axis is precisely in the "past" with respect to O; it is the same space-point as O, only earlier. What happened there then, affects O now. (Unfortunately, that is the way life is.) Another object at Q can get to O by moving with a certain speed less than c, so if this object were in a space ship and moving, it would be, again, the past of the same space-point. That is, in another coordinate system, the axis of time might go through both O and Q. So all points of region 2 are in the "past" of O, and anything that happens in this region *can* affect O. Therefore region 2 is sometimes called the *affective past*, or affecting past; it is the locus of all events which can affect point O in any way.

Region 3, on the other hand, is a region which we can affect *from* O, we can "hit" things by shooting "bullets" out at speeds less than c. So this is the world whose future can be affected by us, and we may call that the *affective future*. Now the interesting thing about all the rest of space-time, i.e., region 1, is that we can neither affect it now *from* O, nor can it affect us now *at* O, because nothing can go faster than the speed of light. Of course, what happens at R *can* affect us *later* that is, if the sun is exploding "right now," it takes eight minutes before we know about it, and it cannot possibly affect us before then.

What we mean by "right now" is a mysterious thing which we cannot define and we cannot affect, but it can affect us later, or we could have affected it if we had done something far enough in the past. When we look at the star Alpha Centauri, we see it as it was four years ago; we might wonder what it is like "now." "Now" means at the same time from our special coordinate system. We can only see Alpha Centauri by the light that has come from our past, up to four years ago, but we do not know what it is doing "now"; it will take four years before what it is doing "now" can affect us. Alpha Centauri "now" is an idea or concept of our mind; it is not something that is really definable physically at the moment, because we have to wait to observe it; we cannot even define it right "now." Furthermore, the "now" depends on the coordinate system. If, for example, Alpha Centauri were moving, an observer there would not agree with us because he would put his axes at an angle, and his "now" would be a *different* time. We have already talked about the fact that simultaneity is not a unique thing.

There are fortune tellers, or people who tell us they can know the future, and there are many wonderful stories about the man who suddenly discovers that he has knowledge about the affective future. Well, there are lots of paradoxes produced by that because if we know something is going to happen, then we can make sure we will avoid it by doing the right thing at the right time, and so on. But actually there is no fortune teller who can even tell us the *present*! There is no one who can tell us what is really happening right now, at any reasonable distance, because that is unobservable. We might ask ourselves this question, which we leave to the student to try to answer: Would any paradox be produced if it were suddenly to become possible to know things that are in the space-like intervals of region 1?

5-4 More about four-vectors

Let us now return to our consideration of the analogy of the Lorentz transformation and rotations of the space axes. We have learned the utility of collecting together other quantities which have the same transformation properties as the coordinates, to form what we call *vectors*, directed lines. In the case of ordinary rotations, there are many quantities that transform the same way as x, y, and z under rotation: for example, the velocity has three components, an x, y, and z component; when seen in a different coordinate system, none of the components is the same, instead they are all transformed to new values. But, somehow or other, the velocity "itself" has a greater reality than do any of its particular components, and we represent it by a directed line.

We therefore ask: Is it or is it not true that there are quantities which transform, or which are related, in a moving system and in a nonmoving system, in the same way as x, y, z, and t? From our experience with vectors, we know that three of the quantities, like x, y, z, would constitute the three components of an ordinary space-vector, but the fourth quantity would look like an ordinary scalar under space rotation, because it does not change so long as we do not go into a moving coordinate system. Is it possible, then, to associate with some of our known "three-vectors" a fourth object, that we could call the "time component," in such a manner that the four objects together would "rotate" the same way as position and time in space-time? We shall now show that there is, indeed, at least one such thing (there are many of them, in fact): *the three components of momentum, and the energy as the time component, transform together* to make what we call a "four-vector." In demonstrating this, since it is quite inconvenient to have to write c's everywhere, we shall use the same trick concerning units of the energy, the mass, and the momentum, that we used in Eq. (5.4). Energy and mass, for example, differ only by a factor c^2 which is merely a question of units, so we can say energy *is* the mass. Instead

Space-Time

of having to write the c^2, we put $E = m$, and then, of course, if there were any trouble we would put in the right amounts of c so that the units would straighten out in the last equation, but not in the intermediate ones.

Thus our equations for energy and momentum are

$$E = m = m_0/\sqrt{1 - v^2},$$
$$\mathbf{p} = m\mathbf{v} = m_0\mathbf{v}/\sqrt{1 - v^2}. \tag{5.6}$$

Also in these units, we have

$$E^2 - p^2 = m_0^2. \tag{5.7}$$

For example, if we measure energy in electron volts, what does a mass of 1 electron volt mean? It means the mass whose rest energy is 1 electron volt, that is, m_0c^2 is one electron volt. For example, the rest mass of an electron is 0.511×10^6 ev.

Now what would the momentum and energy look like in a new coordinate system? To find out, we shall have to transform Eq. (5.6), which we can do because we know how the velocity transforms. Suppose that, as we measure it, an object has a velocity v, but we look upon the same object from the point of view of a space ship which itself is moving with a velocity u, and in that system we use a prime to designate the corresponding thing. In order to simplify things at first, we shall take the case that the velocity v is in the direction of u. (Later, we can do the more general case.) What is v', the velocity as seen from the space ship? It is the composite velocity, the "difference" between v and u. By the law which we worked out before,

$$v' = \frac{v - u}{1 - uv} \, . \tag{5.8}$$

Now let us calculate the new energy E', the energy as the fellow in the space ship would see it. He would use the same rest mass, of course, but he would use v' for the velocity. What we have to do is square v', subtract it from one, take the square root, and take the reciprocal:

$$v'^2 = \frac{v^2 - 2uv + u^2}{1 - 2uv + u^2v^2},$$

$$1 - v'^2 = \frac{1 - 2uv + u^2v^2 - v^2 + 2uv - u^2}{1 - 2uv + u^2v^2}$$

$$= \frac{1 - v^2 - u^2 + u^2v^2}{1 - 2uv + u^2v^2}$$

$$= \frac{(1 - v^2)(1 - u^2)}{(1 - uv)^2}.$$

Therefore

$$\frac{1}{\sqrt{1 - v'^2}} = \frac{1 - uv}{\sqrt{1 - v^2}\sqrt{1 - u^2}}. \tag{5.9}$$

The energy E' is then simply m_0 times the above expression. But we want to express the energy in terms of the unprimed energy and momentum, and we note that

$$E' = \frac{m_0 - m_0uv}{\sqrt{1 - v^2}\sqrt{1 - u^2}} = \frac{(m_0/\sqrt{1 - v^2}) - (m_0v/\sqrt{1 - v^2})u}{\sqrt{1 - u^2}},$$

or

$$E' = \frac{E - up_x}{\sqrt{1 - u^2}}, \tag{5.10}$$

which we recognize as being exactly of the same form as

$$t' = \frac{t - ux}{\sqrt{1 - u^2}}.$$

Next we must find the new momentum p'_x. This is just the energy E times v', and is also simply expressed in terms of E and p:

$$p'_x = E'v' = \frac{m_0(1 - uv)}{\sqrt{1 - v^2}\sqrt{1 - u^2}} \cdot \frac{v - u}{(1 - uv)} = \frac{m_0v - m_0u}{\sqrt{1 - v^2}\sqrt{1 - u^2}}.$$

Thus

$$p'_x = \frac{p_x - uE}{\sqrt{1 - u^2}}, \tag{5.11}$$

Space-Time

which we recognize as being of precisely the same form as

$$x' = \frac{x - ut}{\sqrt{1 - u^2}} \cdot$$

Thus the transformations for the new energy and momentum in terms of the old energy and momentum are exactly the same as the transformations for t' in terms of t and x, and x' in terms of x and t: all we have to do is, every time we see t in (5.4) substitute E, and every time we see x substitute p_x, and then the equations (5.4) will become the same as Eqs. (5.10) and (5.11). This would imply, if everything works right, an additional rule that $p'_y = p_y$ and that $p'_z = p_z$. To prove this would require our going back and studying the case of motion up and down. Actually, we did study the case of motion up and down in the last chapter. We analyzed a complicated collision and we noticed that, in fact, the transverse momentum is *not* changed when viewed from a moving system; so we have already verified that $p'_y = p_y$ and $p'_z = p_z$. The complete transformation, then, is

$$p'_x = \frac{p_x - uE}{\sqrt{1 - u^2}},$$

$$p'_y = p_y,$$

$$p'_z = p_z, \tag{5.12}$$

$$E' = \frac{E - up_x}{\sqrt{1 - u^2}} \cdot$$

In these transformations, therefore, we have discovered four quantities which transform like x, y, z, and t, and which we call the *four-vector momentum*. Since the momentum is a four-vector, it can be represented on a space-time diagram of a moving particle as an "arrow" tangent to the path, as shown in Figure 5-4. This arrow has a time component equal to the energy, and its space components represent its three-vector momentum; this arrow is more "real" than either the energy or the momentum, because those just depend on how we look at the diagram.

Figure 5-4. The four-vector momentum of a particle.

5-5 Four-vector algebra

The notation for four-vectors is different than it is for three-vectors. In the case of three-vectors, if we were to talk about the ordinary three-vector momentum we would write it **p**. If we wanted to be more specific, we could say it has three components which are, for the axes in question, p_x, p_y, and p_z, or we could simply refer to a general component as p_i, and say that i could either be x, y, or z, and that these are the three components; that is, imagine that i is any one of three directions, x, y, or z. The notation that we use for four-vectors is analogous to this: we write p_μ for the four-vector, and μ stands for the *four* possible directions t, x, y, or z.

We could, of course, use any notation we want; do not laugh at notations; invent them, they are powerful. In fact, mathematics is, to a large extent, invention of better notations. The whole idea of a four-vector, in fact, is an improvement in notation so that the transformations can be remembered easily. A_μ, then, is a general four-vector, but for the special case of momentum, the p_t is identified as the energy, p_x is the momentum in the x-direction, p_y is that in the y-direction, and p_z is that in the z-direction. To add four-vectors, we add the corresponding components.

If there is an equation among four-vectors, then the equation is true for *each component*. For instance, if the law of conservation of three-vector momentum is to be true in particle collisions, i.e., if the sum of the momenta for a large number of interacting or colliding particles is to be a constant, that must mean that the sums of all momenta in the x-direction, in the y-direction, and in the z-

Space-Time

direction, for all the particles, must each be constant. This law alone would be impossible in relativity because it is *incomplete*; it is like talking about only two of the components of a three-vector. It is incomplete because if we rotate the axes, we mix the various components, so we must include all three components in our law. Thus, in relativity, we must complete the law of conservation of momentum by extending it to include the *time* component. This is *absolutely necessary* to go with the other three, or there cannot be relativistic invariance. The *conservation of energy* is the fourth equation which goes with the conservation of momentum to make a valid four-vector relationship in the geometry of space and time. Thus the law of conservation of energy and momentum in four-dimensional notation is

$$\sum_{\substack{\text{particles} \\ \text{in}}} p_\mu = \sum_{\substack{\text{particles} \\ \text{out}}} p_\mu, \tag{5.13}$$

or, in a slightly different notation

$$\sum_i p_{i\mu} = \sum_j p_{j\mu}, \tag{5.14}$$

where $i = 1, 2, \ldots$ refers to the particles going into the collision, $j = 1, 2, \ldots$ refers to the particles coming out of the collision, and $\mu = x, y, z,$ or t. You say, "In which axes?" It makes no difference. The law is true for each component, using *any* axes.

In vector analysis we discussed one other thing, the dot product of two vectors. Let us now consider the corresponding thing in space-time. In ordinary rotation we discovered there was an unchanged quantity $x^2 + y^2 + z^2$. In four dimensions, we find that the corresponding quantity is $t^2 - x^2 - y^2 - z^2$ (Eq. 5.3). How can we write that? One way would be to write some kind of four-dimensional thing with a square dot between, like $A_\mu \odot B_\mu$; one of the notations which is actually used is

$$\sum_\mu{}' A_\mu A_\mu = A_t^2 - A_x^2 - A_y^2 - A_z^2. \tag{5.15}$$

The prime on Σ means that the first term, the "time" term, is positive, but the other three terms have minus signs. This quantity, then, will be the same in any coordinate system, and we may call it the square of the length of the four-vector. For instance, what is the square of the length of the four-vector momentum of a single particle? This will be equal to $p_t^2 - p_x^2 - p_y^2 - p_z^2$ or, in other words, $E^2 - p^2$, because we know that p_t is E. What is $E^2 - p^2$? It must be something which is the same in every coordinate system. In particular, it must be the same for a coordinate system which is moving right along with the particle, in which the particle is standing still. If the particle is standing still, it would have no momentum. So in that coordinate system, it is purely its energy, which is the same as its rest mass. Thus $E^2 - p^2 = m_0^2$. So we see that the square of the length of this vector, the four-vector momentum, is equal to m_0^2.

From the square of a vector, we can go on to invent the "dot product," or the product which is a scalar: if a_μ is one four-vector and b_μ is another four-vector, then the scalar product is

$$\sum{}' a_\mu b_\mu = a_t b_t - a_x b_x - a_y b_y - a_z b_z. \qquad (5.16)$$

It is the same in all coordinate systems.

Finally, we shall mention certain things whose rest mass m_0 is zero. A photon of light, for example. A photon is like a particle, in that it carries an energy and a momentum. The energy of a photon is a certain constant, called Planck's constant, times the frequency of the photon: $E = h\nu$. Such a photon also carries a momentum, and the momentum of a photon (or of any other particle, in fact) is h divided by the wavelength: $p = h/\lambda$. But, for a photon, there is a definite relationship between the frequency and the wavelength: $\nu = c/\lambda$. (The number of waves per second, times the wavelength of each, is the distance that the light goes in one second, which, of course, is c.) Thus we see immediately that the energy of a photon must be the momentum times c, or if $c = 1$, *the energy and momentum are equal*. That is to say, the rest mass is zero. Let us look at that

Space-Time

again; that is quite curious. If it is a particle of zero rest mass, what happens when it stops? *It never stops!* It always goes at the speed c. The usual formula for energy is $m_0/\sqrt{1 - v^2}$. Now can we say that $m_0 = 0$ and $v = 1$, so the energy is 0? We *cannot* say that it is zero; the photon really can (and does) have energy even though it has no rest mass, but this it possesses by perpetually going at the speed of light!

We also know that the momentum of any particle is equal to its total energy times its velocity: if $c = 1$, $p = vE$ or, in ordinary units, $p = vE/c^2$. For any particle moving at the speed of light, $p = E$ if $c = 1$. The formulas for the energy of a photon as seen from a moving system are, of course, given by Eq. (5.12), but for the momentum we must substitute the energy times c (or times 1 in this case). The different energies after transformation means that there are different frequencies. This is called the Doppler effect, and one can calculate it easily from Eq. (5.12), using also $E = p$ and $E = hv$.

As Minkowski said, "Space of itself, and time of itself will sink into mere shadows, and only a kind of union between them shall survive."

Six

CURVED SPACE

6-1 Curved spaces with two dimensions

◆ According to Newton, everything attracts everything else with a force inversely proportional to the square of the distance from it, and objects respond to forces with accelerations proportional to the forces. They are Newton's laws of universal gravitation and of motion. As you know, they account for the motions of balls, planets, satellites, galaxies, and so forth.

Einstein had a different interpretation of the law of gravitation. According to him, space and time—which must be put together as space-time—are *curved* near heavy masses. And it is the attempt of things to go along "straight lines" in this curved space-time which makes them move the way they do. Now that is a complex idea—very complex. It is the idea we want to explain in this chapter.

Our subject has three parts. One involves the effects of gravitation. Another involves the ideas of space-time which we already studied. The third involves the idea of curved space-time. We will simplify our subject in the beginning by not worrying about gravity and by leaving out the time—discussing just curved space. We will talk later about the other parts, but we will concentrate now on the idea of curved space—what is meant by curved space, and, more specifically, what is meant by curved space in this application of Einstein. Now even that much turns out to be somewhat difficult in

111

three dimensions. So we will first reduce the problem still further and talk about what is meant by the words "curved space" in two dimensions.

In order to understand this idea of curved space in two dimensions you really have to appreciate the limited point of view of the character who lives in such a space. Suppose we imagine a bug with no eyes who lives on a plane, as shown in Figure 6-1. He can move only on the plane, and he has no way of knowing that there is any way to discover any "outside world." (He hasn't got your imagination.) We are, of course, going to argue by analogy. *We* live in a three-dimensional world, and we don't have any imagination about going off our three-dimensional world in a new direction; so we have to think the thing out by analogy. It is as though we were bugs living on a plane, and there was a space in another direction. That's why we will first work with the bug, remembering that he must live on his surface and can't get out.

As another example of a bug living in two dimensions, let's imagine one who lives on a sphere. We imagine that he can walk around on the surface of the sphere, as in Figure 6-2, but that he can't look "up," or "down," or "out."

Now we want to consider still a *third* kind of creature. He is also a bug like the others, and also lives on a plane, as our first bug did, but this time the plane is peculiar. The temperature is different at different places. Also, the bug and any rulers he uses are all made of

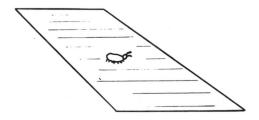

Figure 6-1. A bug on a plane surface.

Curved Space

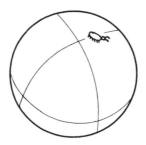

Figure 6-2. A bug on a sphere.

the same material which expands when it is heated. Whenever he puts a ruler somewhere to measure something the ruler expands immediately to the proper length for the temperature at that place. Wherever he puts any object—himself, a ruler, a triangle, or anything—the thing stretches itself because of the thermal expansion. Everything is longer in the hot places than it is in the cold places, and everything has the same coefficient of expansion. We will call the home of our third bug a "hot plate," although we will particularly want to think of a special kind of hot plate that is cold in the center and gets hotter as we go out toward the edges (Figure 6-3).

Now we are going to imagine that our bugs begin to study geometry. Although we imagine that they are blind so that they can't see any "outside" world, they can do a lot with their legs and feelers.

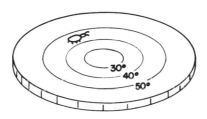

Figure 6-3. A bug on a hot plate.

SIX NOT-SO-EASY PIECES

They can draw lines, and they can make rulers, and measure off lengths. First, let's suppose that they start with the simplest idea in geometry. They learn how to make a straight line—defined as the shortest line between two points. Our first bug—see Figure 6-4—learns to make very good lines. But what happens to the bug on the sphere? He draws his straight line as the shortest distance—*for him*—between two points, as in Figure 6-5. It may look like a curve to us, but he has no way of getting off the sphere and finding out that there is "really" a shorter line. He just knows that if he tries any other path *in his world* it is always longer than his straight line. So we will let him have his straight line as the shortest arc between two points. (It is, of course an arc of a great circle.)

Finally, our third bug—the one in Figure 6-3—will also draw "straight lines" that look like curves to us. For instance, the shortest distance between *A* and *B* in Figure 6-6 would be on a curve like the one shown. Why? Because when his line curves out toward the warmer parts of his hot plate, the rulers get longer (from our omniscient point of view) and it takes fewer "yardsticks" laid end-to-end to get from *A* to *B*. So *for him* the line is straight—he has no way of knowing that there could be someone out in a strange three-dimensional world who would call a different line "straight."

We think you get the idea now that all the rest of the analysis will always be from the point of view of the creatures on the particular surfaces and not from *our* point of view. With that in mind let's see what the rest of their geometries looks like. Let's assume that the

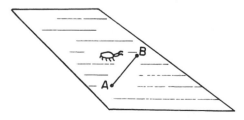

Figure 6-4. Making a "straight line" on a plane.

Curved Space

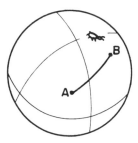

Figure 6-5. Making a "straight line" on a sphere.

bugs have all learned how to make two lines intersect at right angles. (You can figure out how they could do it.) Then our first bug (the one on the normal plane) finds an interesting fact. If he starts at the point *A* and makes a line 100 inches long, then makes a right angle and marks off another 100 inches, then makes another right angle and goes another 100 inches, then makes a third right angle and a fourth line 100 inches long, he ends up right at the starting point as shown in Figure 6-7(a). It is a property of his world—one of the facts of his "geometry."

Then he discovers another interesting thing. If he makes a triangle—a figure with three straight lines—the sum of the angles is equal to 180°, that is, to the sum of two right angles. See Figure 6-7(b).

Then he invents the circle. What's a circle? A circle is made this way: You rush off on straight lines in many many directions from a

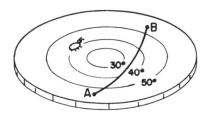

Figure 6-6. Making a "straight line" on the hot plate.

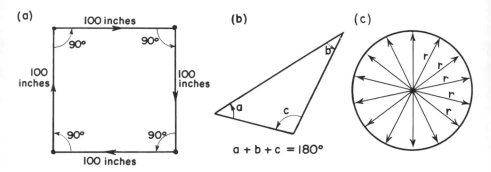

Figure 6-7. A square, triangle, and circle in a flat space.

single point, and lay out a lot of dots that are all the same distance from that point. See Figure 6-7(c). (We have to be careful how we define these things because we've got to be able to make the analogs for the other fellows.) Of course, it's equivalent to the curve you can make by swinging a ruler around a point. Anyway, our bug learns how to make circles. Then one day he thinks of measuring the distance around a circle. He measures several circles and finds a neat relationship: The distance around is always the same number times the radius r (which is, of course, the distance from the center out to the curve). The circumference and the radius always have the same ratio—approximately 6.283—independent of the size of the circle.

Now let's see what our other bugs have been finding out about *their* geometries. First, what happens to the bug on the sphere when he tries to make a "square"? If he follows the prescription we gave above, he would probably think that the result was hardly worth the trouble. He gets a figure like the one shown in Figure 6-8. His endpoint B isn't on top of the starting point A. It doesn't work out to a closed figure at all. Get a sphere and try it. A similar thing would happen to our friend on the hot plate. If he lays out four straight lines of equal length—as measured with his expanding rulers—joined by right angles he gets a picture like the one in Figure 6-9.

Curved Space

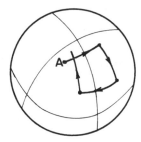

Figure 6-8. Trying to make a "square" on a sphere.

Now suppose that our bugs had each had their own Euclid who had told them what geometry "should" be like, and that they had checked him out roughly by making crude measurements on a *small* scale. Then as they tried to make accurate squares on a larger scale they would discover that something was wrong. The point is, that just by *geometrical measurements* they would discover that something was the matter with their space. We define a *curved space* to be a space in which the geometry is not what we expect for a plane. The geometry of the bugs on the sphere or on the hot plate is the geometry of a curved space. The rules of Euclidian geometry fail. And it isn't necessary to be able to lift yourself out of the plane in order to find out that the world that you live in is curved. It isn't necessary to circumnavigate the globe in order to find out that it is a ball. You can find out that you live

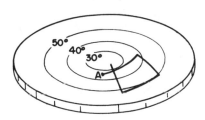

Figure 6-9. Trying to make a "square" on the hot plate.

SIX NOT SO-EASY PIECES

on a ball by laying out a square. If the square is very small you will need a lot of accuracy, but if the square is large the measurement can be done more crudely.

Let's take the case of a triangle on a plane. The sum of the angles is 180 degrees. Our friend on the sphere can find triangles that are very peculiar. He can, for example, find triangles which have *three right angles*. Yes indeed! One is shown in Figure 6-10. Suppose our bug starts at the north pole and makes a straight line all the way down to the equator. Then he makes a right angle and another perfect straight line the same length. Then he does it again. For the very special length he has chosen he gets right back to his starting point, and also meets the first line with a right angle. So there is no doubt that for him this triangle has three right angles, or 270 degrees in the sum. It turns out that for him the sum of the angles of the triangle is *always* greater than 180 degrees. In fact, the excess (for the special case shown, the extra 90 degrees) is proportional to how much area the triangle has. If a triangle on a sphere is very small, its angles add up to very nearly 180 degrees, only a little bit over. As the triangle gets bigger the discrepancy goes up. The bug on the hot plate would discover similar difficulties with his triangles.

Let's look next at what our other bugs find out about circles. They make circles and measure their circumferences. For example,

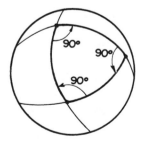

Figure 6-10. On a sphere a "triangle" can have three 90° angles.

Curved Space

the bug on the sphere might make a circle like the one shown in Figure 6-11. And he would discover that the circumference is *less* than 2π times the radius. (You can see that because from the wisdom of our three-dimensional view it is obvious that what he calls the "radius" is a curve which is *longer* than the true radius of the circle.) Suppose that the bug on the sphere had read Euclid, and decided to predict a radius by dividing the circumference C by 2π, taking

$$r_{\text{pred}} = \frac{C}{2\pi}. \qquad (6.1)$$

Then he would find that the measured radius was larger than the predicted radius. Pursuing the subject, he might define the difference to be the "excess radius," and write

$$r_{\text{meas}} - r_{\text{pred}} = r_{\text{excess}}, \qquad (6.2)$$

and study how the excess radius effect depended on the size of the circle.

Our bug on the hot plate would discover a similar phenomenon. Suppose he was to draw a circle centered at the cold spot on the plate as in Figure 6-12. If we were to watch him as he makes the circle we would notice that his rulers are short near the center and get longer as they are moved outward—although the bug doesn't know it, of course. When he measures the circumference the ruler is

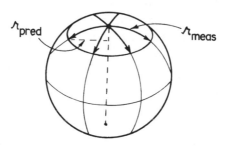

Figure 6-11. Making a circle on a sphere.

SIX NOT-SO-EASY PIECES

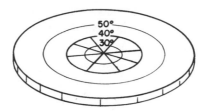

Figure 6-12. Making a circle on the hot plate.

long all the time, so he, too, finds out that the measured radius is longer than the predicted radius, $C/2\pi$. The hot-plate bug also finds an "excess radius effect." And again the size of the effect depends on the radius of the circle.

We will *define* a "curved space" as one in which these types of geometrical errors occur: The sum of the angles of a triangle is different from 180 degrees; the circumference of a circle divided by 2π is not equal to the radius; the rule for making a square doesn't give a closed figure. You can think of others.

We have given two different examples of curved space: the sphere and the hot plate. But it is interesting that if we choose the right temperature variation as a function of distance on the hot plate, the two *geometries* will be exactly the same. It is rather amusing. We can make the bug on the hot plate get exactly the same answers as the bug on the ball. For those who like geometry and geometrical problems we'll tell you how it can be done. If you assume that the length of the rulers (as determined by the temperature) goes in proportion to one plus some constant times the square of the distance away from the origin, then you will find that the geometry of that hot plate is exactly the same in all details* as the geometry of the sphere.

There are, of course, other kinds of geometry. We could ask about the geometry of a bug who lived on a pear, namely something which has a sharper curvature in one place and a weaker curvature

---

* except for the one point at infinity.

Curved Space

in the other place, so that the excess in angles in triangles is more severe when he makes little triangles in one part of his world than when he makes them in another part. In other words, the curvature of a space can vary from place to place. That's just a generalization of the idea. It can also be imitated by a suitable distribution of temperature on a hot plate.

We may also point out that the results could come out with the opposite kind of discrepancies. You could find out, for example, that all triangles when they are made too large have the sum of their angles *less* than 180 degrees. That may sound impossible, but it isn't at all. First of all, we could have a hot plate with the temperature decreasing with the distance from the center. Then all the effects would be reversed. But we can also do it purely geometrically by looking at the two-dimensional geometry of the surface of a saddle. Imagine a saddle-shaped surface like the one sketched in Figure 6-13. Now draw a "circle" on the surface, defined as the locus of all points the same distance from a center. This circle is a curve that oscillates up and down with a scallop effect. So its circumference is larger than you would expect from calculating $2\pi r$. So $C/2\pi$ is now less than r. The "excess radius" would be negative.

Spheres and pears and such are all surfaces of *positive* curvatures; and the others are called surfaces of *negative* curvature. In general, a two-dimensional world will have a curvature which varies from place to place and may be positive in some places and negative in

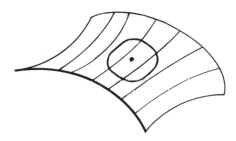

Figure 6-13. A "circle" on a saddle-shaped surface.

other places. In general, we mean by a curved space simply one in which the rules of Euclidean geometry break down with one sign of discrepancy or the other. The amount of curvature—defined, say, by the excess radius—may vary from place to place.

We might point out that, from our definition of curvature, a cylinder is, surprisingly enough, not curved. If a bug lived on a cylinder, as shown in Figure 6-14, he would find out that triangles, squares, and circles would all have the same behavior they have on a plane. This is easy to see, by just thinking about how all the figures will look if the cylinder is unrolled onto a plane. Then all the geometrical figures can be made to correspond exactly to the way they are in a plane. So there is no way for a bug living on a cylinder (assuming that he doesn't go all the way around, but just makes local measurements) to discover that his space is curved. In our technical sense, then, we consider that his space is *not* curved. What we want to talk about is more precisely called *intrinsic* curvature; that is, a curvature which can be found by measurements only in a local region. (A cylinder has no intrinsic curvature.) This was the

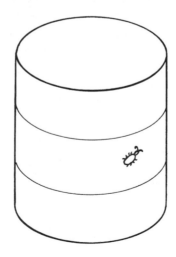

Figure 6-14. A two-dimensional space with zero intrinsic curvature.

sense intended by Einstein when he said that our space is curved. But we as yet only have defined a curved space in two dimensions; we must go onward to see what the idea might mean in three dimensions.

6-2 *Curvature in three-dimensional space*

We live in three-dimensional space and we are going to consider the idea that three-dimensional space is curved. You say, "But how can you imagine it being bent in any direction?" Well, we can't imagine space being bent in any direction because our imagination isn't good enough. (Perhaps it's just as well that we can't imagine too much, so that we don't get too free of the real world.) But we can still *define* a curvature without getting out of our three-dimensional world. All we have been talking about in two dimensions was simply an exercise to show how we could get a definition of curvature which didn't require that we be able to "look in" from the outside.

We can determine whether our world is curved or not in a way quite analogous to the one used by the gentlemen who live on the sphere and on the hot plate. We may not be able to distinguish between two such cases but we certainly can distinguish those cases from the flat space, the ordinary plane. How? Easy enough: We lay out a triangle and measure the angles. Or we make a great big circle and measure the circumference and the radius. Or we try to lay out some accurate squares, or try to make a cube. In each case we test whether the laws of geometry work. If they don't work, we say that our space is curved. If we lay out a big triangle and the sum of its angles exceeds 180 degrees, we can say our space is curved. Or if the measured radius of a circle is not equal to its circumference over 2π, we can say our space is curved.

You will notice that in three dimensions the situation can be much more complicated than in two. At any one place in two dimensions there is a certain amount of curvature. But in three dimensions there can be *several components* to the curvature. If we lay out a triangle in some plane, we may get a different answer than if we orient the plane of the triangle in a different way. Or take the example of a circle. Suppose we draw a circle and measure the radius and it doesn't check with $C/2\pi$ so that there is some excess radius. Now we draw another circle at right angles—as in Figure 6-15. There's no need for the excess to be exactly the same for both circles. In fact, there might be a positive excess for a circle in one plane, and a defect (negative excess) for a circle in the other plane.

Perhaps you are thinking of a better idea: Can't we get around all of these components by using a *sphere* in three dimensions? We can specify a sphere by taking all the points that are the same distance from a given point in space. Then we can measure the surface area by laying out a fine scale rectangular grid on the surface of the sphere and adding up all the bits of area. According to Euclid the total area A is supposed to be 4π times the square of the radius; so we can define a "predicted radius" as $\sqrt{A/4\pi}$. But we can also measure the radius directly by digging a hole to the center and measuring the distance. Again, we can take the measured radius minus the predicted radius and call the difference the radius excess.

$$r_{excess} = r_{meas} - \left(\frac{\text{measured area}}{4\pi}\right)^{1/2},$$

which would be a perfectly satisfactory measure of the curvature.

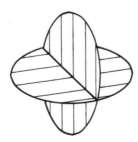

Figure 6-15. The excess radius may be different for circles with different orientations.

Curved Space

It has the great advantage that it doesn't depend upon how we orient a triangle or a circle.

But the excess radius of a sphere also has a disadvantage; it doesn't completely characterize the space. It gives what is called the *mean curvature* of the three-dimensional world, since there is an averaging effect over the various curvatures. Since it is an average, however, it does not solve completely the problem of defining the geometry. If you know only this number you can't predict all properties of the geometry of the space, because you can't tell what would happen with circles of different orientation. The complete definition requires the specification of six "curvature numbers" at each point. Of course the mathematicians know how to write all those numbers. You can read someday in a mathematics book how to write them all in a high-class and elegant form, but it is first a good idea to know in a rough way what it is that you are trying to write about. For most of our purposes the average curvature will be enough.*

6-3 Our space is curved

Now comes the main question. Is it true? That is, is the actual physical three-dimensional space we live in curved? Once we have enough imagination to realize the possibility that space might be curved, the human mind naturally gets curious about whether the real world is curved or not. People have made direct geometrical measurements to try to find out, and haven't found any deviations.

---

* We should mention one additional point for completeness. If you want to carry the hot-plate model of curved space over into three dimensions you must imagine that the length of the ruler depends not only on where you put it, but also on which orientation the ruler has when it is laid down. It is a generalization of the simple case in which the length of the ruler depends on where it is, but is the same if set north-south, or east-west, or up-down. This generalization is needed if you want to represent a three-dimensional space with any arbitrary geometry with such a model, although it happens not to be necessary for two dimensions.

On the other hand, by arguments about gravitation, Einstein discovered that space *is* curved, and we'd like to tell you what Einstein's law is for the amount of curvature, and also tell you a little bit about how he found out about it.

Einstein said that space is curved and that matter is the source of the curvature. (Matter is also the source of gravitation, so gravity is related to the curvature—but that will come later in the chapter.) Let us suppose, to make things a little easier, that the matter is distributed continuously with some density, which may vary, however, as much as you want from place to place.* The rule that Einstein gave for the curvature is the following: If there is a region of space with matter in it and we take a sphere small enough that the density ρ of matter inside it is effectively constant, then the *radius excess* for the sphere is proportional to the mass inside the sphere. Using the definition of excess radius, we have

$$\text{Radius excess} = \sqrt{\frac{A}{4\pi}} - r_{\text{meas}} = \frac{G}{3c^2} \cdot M. \qquad (6.3)$$

Here, G is the gravitational constant (of Newton's theory), c is the velocity of light, and $M = 4\pi\rho r^3/3$ is the mass of the matter inside the sphere. This is Einstein's law for the mean curvature of space.

Suppose we take the earth as an example and forget that the density varies from point to point—so we won't have to do any integrals. Suppose we were to measure the surface of the earth very carefully, and then dig a hole to the center and measure the radius. From the surface area we could calculate the predicted radius we would get from setting the area equal to $4\pi r^2$. When we compared the predicted radius with the actual radius, we would find that the actual radius exceeded the predicted radius by the amount given in Eq. (6.3). The constant $G/3c^2$ is about 2.5×10^{-29} cm per gram, so for each gram of material the measured radius is off by 2.5×10^{-29} cm. Putting in the mass of the earth, which is about 6×10^{27} grams, it

---

* Nobody—not even Einstein—knows how to do it if mass comes concentrated at points.

Curved Space

turns out that the earth has 1.5 millimeters more radius than it should have for its surface area.* Doing the same calculation for the sun, you find that the sun's radius is one-half a kilometer too long.

You should note that the law says that the *average* curvature *above* the surface area of the earth is zero. But that does *not* mean that all the components of the curvature are zero. There may still be—and, in fact, there is—some curvature above the earth. For a circle in a plane there will be an excess radius of one sign for some orientations and of the opposite sign for other orientations. It just turns out that the average over a sphere is zero when there is no mass *inside* it. Incidentally, it turns out that there is a relation between the various components of the curvature and the *variation* of the average curvature from place to place. So if you know the average curvature everywhere, you can figure out the details of the curvature at each place. The average curvature above the earth varies with altitude, so the space there is curved. And it is that curvature that we see as a gravitational force.

Suppose we have a bug on a plane, and suppose that the "plane" has little pimples in the surface. Wherever there is a pimple the bug would conclude that his space has little local regions of curvature. We have the same thing in three dimensions. Wherever there is a lump of matter, our three-dimensional space has a local curvature—a kind of three-dimensional pimple.

If we make a lot of bumps on a plane there might be an overall curvature besides all the pimples—the surface might become like a ball. It would be interesting to know whether our space has a net average curvature as well as the local pimples due to the lumps of matter like the earth and the sun. The astrophysicists have been trying to answer that question by making measurements of galaxies at very large distances. For example, if the number of galaxies we see in a spherical shell at a large distance is different from what we

---

* Approximately, because the density is not independent of radius as we are assuming.

would expect from our knowledge of the radius of the shell, we would have a measure of the excess radius of a tremendously large sphere. From such measurements it is hoped to find out whether our whole universe is flat on the average, or round—whether it is "closed," like a sphere, or "open" like a plane. You may have heard about the debates that are going on about this subject. There are debates because the astronomical measurements are still completely inconclusive; the experimental data are not precise enough to give a definite answer. Unfortunately, we don't have the slightest idea about the overall curvature of our universe on a large scale.

6-4 Geometry in space-time

Now we have to talk about time. As you know from the special theory of relativity, measurements of space and measurements of time are interrelated. And it would be kind of crazy to have something happening to the space, without the time being involved in the same thing. You will remember that the measurement of time depends on the speed at which you move. For instance, if we watch a guy going by in a space ship we see that things happen more slowly for him than for us. Let's say he takes off on a trip and returns in 100 seconds flat *by our watches*; his watch might say that he had been gone for only 95 seconds. In comparison with ours, his watch—and all other processes, like his heart beat—have been running slow.

Now let's consider an interesting problem. Suppose you are the one in the space ship. We ask you to start off at a given signal and return to your starting place just in time to catch a later signal—at, say, exactly 100 seconds later according to *our* clock. And you are also asked to make the trip in such a way that *your* watch will show the *longest possible* elapsed time. How should you move? You should stand still. If you move at all your watch will read less than 100 seconds when you get back.

Curved Space

Suppose, however, we change the problem a little. Suppose we ask you to start at Point A on a given signal and go to point B (both fixed relative to us), and to do it in such a way that you arrive back just at the time of a second signal (say 100 seconds later according to our fixed clock). Again you are asked to make the trip in the way that lets you arrive with the latest possible reading on your watch. How would you do it? For which path and schedule will *your* watch show the greatest elapsed time when you arrive? The answer is that you will spend the longest time from *your* point of view if you make the trip by going at a uniform speed along a straight line. Reason: Any extra motions and any extra-high speeds will make your clock go slower. (Since the time deviations depend on the *square* of the velocity, what you lose by going extra fast at one place you can never make up by going extra slowly in another place.)

The point of all this is that we can use the idea to define "a straight line" in space-time. The analog of a straight line in space is for space-time a *motion* at uniform velocity in a constant direction.

The curve of shortest distance in space corresponds in space-time not to the path of shortest time, but to the one of *longest* time, because of the funny things that happen to signs of the t-terms in relativity. "Straight-line" motion—the analog of "uniform velocity along a straight line"—is then that motion which takes a watch from one place at one time to another place at another time in the way that gives the longest time reading for the watch. This will be our definition for the analog of a straight line in space-time.

6-5 Gravity and the principle of equivalence

Now we are ready to discuss the laws of gravitation. Einstein was trying to generate a theory of gravitation that would fit with the relativity theory that he had developed earlier. He was struggling along until he latched onto one important principle which guided him into getting the correct laws. That principle is based on the idea that when a thing is falling freely everything inside it seems weightless. For example, a satellite in orbit is falling freely in the earth's gravity, and an astronaut in it feels weightless. This idea, when stated with greater precision, is called *Einstein's principle of equivalence*. It depends on the fact that all objects fall with exactly the same acceleration no matter what their mass, or what they are made of. If we have a spaceship that is "coasting"—so it's in a free fall—and there is a man inside, then the laws governing the fall of the man and the ship are the same. So if he puts himself in the middle of the ship he will stay there. He doesn't fall *with respect to the ship*. That's what we mean when we say he is "weightless."

Now suppose you are in a rocket ship which is accelerating. Accelerating with respect to what? Let's just say that its engines are on and generating a thrust so that it is not coasting in a free fall. Also imagine that you are way out in empty space so that there are practically no gravitational forces on the ship. If the ship is accelerating with "lg" you will be able to stand on the "floor" and will feel your normal weight. Also if you let go of a ball, it will "fall" toward the floor. Why? Because the ship is accelerating "upward," but the ball has no forces on it, so it will not accelerate; it will get left behind. Inside the ship the ball will appear to have a downward acceleration of "lg."

Now let's compare that with the situation in a spaceship sitting at rest on the surface of the earth. *Everything is the same!* You would be pressed toward the floor, a ball would fall with an acceleration of lg, and so on. In fact, how could you tell inside a space ship whether you are sitting on the earth or are accelerating in free space? According to Einstein's equivalence principle there is no way to tell if you only make measurements of what happens to things inside!

To be strictly correct, that is true only for one point inside the ship. The gravitational field of the earth is not precisely uniform, so a freely falling ball has a slightly different acceleration at different places—the direction changes and the magnitude changes. But if we imagine a strictly uniform gravitational field, it is completely imitated in every respect by a system with a constant acceleration. That is the basis of the principle of equivalence.

Curved Space

6-6 *The speed of clocks in a gravitational field*

Now we want to use the principle of equivalence for figuring out a strange thing that happens in a gravitational field. We'll show you something that happens in a rocket ship which you probably wouldn't have expected to happen in a gravitational field. Suppose we put a clock at the "head" of the rocket ship—that is, at the "front" end—and we put another identical clock at the "tail," as in Figure 6-16. Let's call the two clocks A and B. If we compare these two clocks when the ship is accelerating, the clock at the head seems

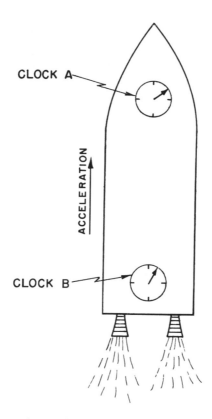

Figure 6-16. An accelerating rocket ship with two clocks.

to run fast relative to the one at the tail. To see that, imagine that the front clock emits a flash of light each second, and that you are sitting at the tail comparing the arrival of the light flashes with the ticks of clock B. Let's say that the rocket is in the position a of Figure 6-17 when clock A emits a flash, and at the position b when the flash arrives at clock B. Later on the ship will be at position c when the clock A emits its next flash, and at position d when you see it arrive at clock B.

The first flash travels the distance L_1 and the second flash travels the shorter distance L_2. It is a shorter distance because the ship is accelerating and has a higher speed at the time of the second flash. You can see, then, that if the two flashes were emitted from clock A one second apart, they would arrive at clock B with a separation somewhat less than one second, since the second flash doesn't spend as much time on the way. The same thing will also happen for all the later flashes. So if you were sitting in the tail you would conclude that clock A was running faster than clock B. If you were to do the same thing in reverse—letting clock B emit light and observing it at clock A—you would conclude that B was running

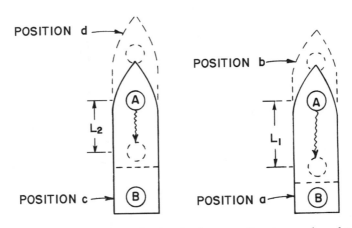

Figure 6-17. A clock at the head of an accelerating rocket ship appears to run faster than a clock at the tail.

slower than *A*. Everything fits together and there is nothing mysterious about it all.

But now let's think of the rocket ship at rest in the earth's gravity. *The same thing happens.* If you sit on the floor with one clock and watch another one which is sitting on a high shelf, it will appear to run faster than the one on the floor! You say, "But that is wrong. The times should be the same. With no acceleration there's no reason for the clocks to appear to be out of step." But they must if the principle of equivalence is right. And Einstein insisted that the principle *was* right, and went courageously and correctly ahead. He proposed that clocks at different places in a gravitational field must appear to run at different speeds. But if one always *appears* to be running at a different speed with respect to the other, then so far as the first is concerned the other *is* running at a different rate.

But now you see we have the analog for clocks of the hot ruler we were talking about earlier, when we had the bug on a hot plate. We imagined that rulers and bugs and everything changed lengths in the same way at various temperatures so they could never tell that their measuring sticks were changing as they moved around on the hot plate. It's the same with clocks in a gravitational field. Every clock we put at a higher level is seen to go faster. Heartbeats go faster, all processes run faster.

If they didn't you would be able to tell the difference between a gravitational field and an accelerating reference system. The idea that time can vary from place to place is a difficult one, but it is the idea Einstein used, and it is correct—believe it or not.

Using the principle of equivalence we can figure out how much the speed of a clock changes with height in a gravitational field. We just work out the apparent discrepancy between the two clocks in the accelerating rocket ship. The easiest way to do this is to use the result we found in Chapter 34 of Vol. I* for the Doppler effect. There, we found—see Eq. (34.14)*—that if v is the *relative* velocity

---

* of the original *Lectures on Physics.*

of a source and a receiver, the *received* frequency ω is related to the *emitted* frequency ω_0 by

$$\omega = \omega_0 \frac{1 + v/c}{\sqrt{1 - v^2/c^2}}. \tag{6.4}$$

Now if we think of the accelerating rocket ship in Figure 6-17 the emitter and receiver are moving with equal velocities at any one instant. But in the time that it takes the light signals to go from clock A to clock B the ship has accelerated. It has, in fact, picked up the additional velocity gt, where g is the acceleration and t is time it takes light to travel the distance H from A to B. This time is very nearly H/c. So when the signals arrive at B, the ship has increased its velocity by gH/c. The receiver always has this velocity *with respect to the emitter* at the instant the signal left it. So this is the velocity we should use in the Doppler shift formula, Eq. (6.4). Assuming that the acceleration and the length of the ship are small enough that this velocity is much smaller than c, we can neglect the term in v^2/c^2. We have that

$$\omega = \omega_0 \left(1 + \frac{gH}{c^2}\right). \tag{6.5}$$

So for the two clocks in the space ship we have the relation

$$(\text{Rate at the receiver}) = (\text{Rate of emission}) \left(1 + \frac{gH}{c^2}\right), \tag{6.6}$$

where H is the height of the emitter *above* the receiver.

From the equivalence principle the same result must hold for two clocks separated by the height H in a gravitational field with the free fall acceleration g.

This is such an important idea we would like to demonstrate that it also follows from another law of physics—from the conservation of energy. We know that the gravitational force on an object is proportional to its mass M, which is related to its total internal energy E by $M = E/c^2$. For instance, the masses of nuclei determined from the *energies* of nuclear reactions which transmute one nucleus into another agree with the masses obtained from atomic *weights*.

Curved Space

Now think of an atom which has a lowest energy state of total energy E_0 and a higher energy state E_1, and which can go from the state E_1 to the state E_0 by emitting light. The frequency ω of the light will be given by

$$\hbar\omega = E_1 - E_0. \tag{6.7}$$

Now suppose we have such an atom in the state E_1 sitting on the floor, and we carry it from the floor to the height H. To do that we must do some work in carrying the mass $m_1 = E_1/c^2$ up against the gravitational force. The amount of work done is

$$\frac{E_1}{c^2} gH. \tag{6.8}$$

Then we let the atom emit a photon and go into the lower energy state E_0. Afterward we carry the atom back to the floor. On the return trip the mass is E_0/c^2; we get back the energy

$$\frac{E_0}{c^2} gH, \tag{6.9}$$

so we have done a net amount of work equal to

$$\Delta U = \frac{E_1 - E_0}{c^2} gH. \tag{6.10}$$

When the atom emitted the photon it gave up the energy $E_1 - E_0$. Now suppose that the photon happened to go down to the floor and be absorbed. How much energy would it deliver there? You might at first think that it would deliver just the energy $E_1 - E_0$. But that can't be right if energy is conserved, as you can see from the following argument. We started with the energy E_1 at the floor. When we finish, the energy at the floor level is the energy E_0 of the atom in its lower state plus the energy E_{ph} received from the photon. In the meantime we have had to supply the additional energy ΔU of Eq. (6.10). If energy is conserved, the energy we end up with at the floor must be greater than we started with by just the work we have done. Namely, we must have that

$$E_{ph} + E_0 = E_1 + \Delta U,$$

or $\qquad\qquad\qquad\qquad\qquad\qquad\qquad\qquad\qquad$ (6.11)

$$E_{ph} = (E_1 - E_0) + \Delta U.$$

It must be that the photon does *not* arrive at the floor with just the energy $E_1 - E_0$ it started with, but with a *little more energy*. Otherwise some energy would have been lost. If we substitute in Eq. (6.11) the ΔU we got in Eq. (6.10), we get that the photon arrives at the floor with the energy

$$E_{ph} = (E_1 - E_0) \left(1 + \frac{gH}{c^2}\right). \qquad (6.12)$$

But a photon of energy E_{ph} has the frequency $\omega = E_{ph}/\hbar$. Calling the frequency of the *emitted* photon ω_0—which is by Eq. (6.7) equal to $(E_1 - E_0)/\hbar$—our result in Eq. (6.12) gives again the relation of (6.5) between the frequency of the photon when it is absorbed on the floor and the frequency with which it was emitted.

The same result can be obtained in still another way. A photon of frequency ω_0 has the energy $E_0 = \hbar\omega_0$. Since the energy E_0 has the gravitational mass E_0/c^2 the photon has a mass (*not* rest mass) $\hbar\omega_0/c^2$, and is "attracted" by the earth. In falling the distance H it will gain an additional energy $(\hbar\omega_0/c^2)gH$, so it arrives with the energy

$$E = \hbar w_0 \left(1 + \frac{gH}{c^2}\right).$$

But its frequency after the fall is $E/\hbar$, giving again the result in Eq. (6.5). Our ideas about relativity, quantum physics, and energy conservation all fit together only if Einstein's predictions about clocks in a gravitational field are right. The frequency changes we are talking about are normally very small. For instance, for an altitude difference of 20 meters at the earth's surface the frequency difference is only about two parts in 10^{15}. However, just such a change has recently been found experimentally using the Mössbauer effect.* Einstein was perfectly correct.

---

* R. V. Pound and G. A. Rebka, Jr., *Physical Review Letters* Vol. 4, p. 337 (1960).

Curved Space

6-7 *The curvature of space-time*

Now we want to relate what we have just been talking about to the idea of curved space-time. We have already pointed out that if the time goes at different rates in different places, it is analogous to the curved space of the hot plate. But it is more than an analogy; it means that space-time *is* curved. Let's try to do some geometry in space-time. That may at first sound peculiar, but we have often made diagrams of space-time with distance plotted along one axis and time along the other. Suppose we try to make a rectangle in space-time. We begin by plotting a graph of height H versus t as in Figure 6-18(a). To make the base of our rectangle we take an object which is *at rest* at the height H_1 and follow its world line for 100 seconds. We get the line BD in part (b) of the figure which is parallel to the t-axis. Now let's take another object which is 100 feet above the first one at $t = 0$. It starts at the point A in Figure 6-18(c). Now we follow its world line for 100 seconds as measured by a clock at A. The object goes from A to C, as shown in part (d) of the figure. But notice that since time goes at a different rate at the two heights—we are assuming that there is a gravitational field—the two points C and D are not simultaneous. If we try to complete the square by drawing a line to the point C' which is 100 feet above D at the same time, as in Figure 6-18(e), the pieces don't fit. And that's what we mean when we say that space-time is curved.

6-8 *Motion in curved space-time*

Let's consider an interesting little puzzle. We have two identical clocks, A and B, sitting together on the surface of the earth as in Figure 6-19. Now we lift clock A to some height H, hold it there awhile, and return it to the ground so that it arrives at just the instant when clock B has advanced by 100 seconds. Then clock A will read something like 107 seconds, because it was running faster when it was up in the air. Now here is the puzzle. How should we

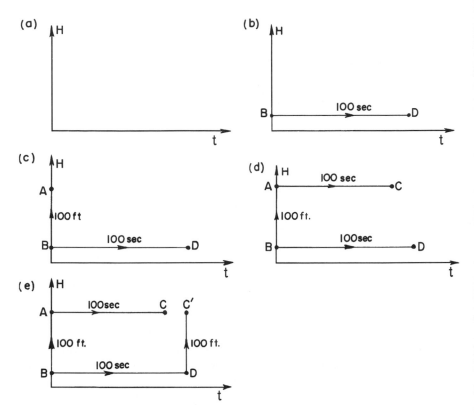

Figure 6-18. Trying to make a rectangle in space-time.

move clock A so that it reads the latest possible time—always assuming that it returns when B reads 100 seconds? You say, "That's easy. Just take A as high as you can. Then it will run as fast as possible, and be the latest when you return." Wrong. You forgot something—we've only got 100 seconds to go up and back. If we go very high, we have to go very fast to get there and back in 100 seconds. And you mustn't forget the effect of special relativity which causes moving clocks to *slow down* by the factor $\sqrt{1 - v^2/c^2}$. This relativity effect works in the direction of making clock A read

Curved Space

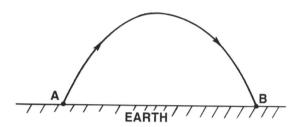

Figure 6-19. In a uniform gravitational field the trajectory with the maximum proper time for a fixed elapsed time is a parabola.

less time than clock B. You see that we have a kind of game. If we stand still with clock A we get 100 seconds. If we go up slowly to a small height and come down slowly we can get a little more than 100 seconds. If we go a little higher, maybe we can gain a little more. But if we go too high we have to move fast to get there, and we may slow down the clock enough that we end up with less than 100 seconds. What program of height versus time—how high to go and with what speed to get there, carefully adjusted to bring us back to clock B when it has increased by 100 seconds—will give us the largest possible time reading on clock A?

Answer: Find out how fast you have to throw a ball up into the air so that it will fall back to earth in exactly 100 seconds. The ball's motion—rising fast, slowing down, stopping, and coming back down—is exactly the right motion to make the time the maximum on a wrist watch strapped to the ball.

Now consider a slightly different game. We have two points A and B both on the earth's surface at some distance from one another. We play the same game that we did earlier to find what we call the straight line. We ask how we should go from A to B so that the time on our moving watch will be the longest—assuming we start at A on a given signal and arrive at B on another signal at B which we will say is 100 seconds later by a fixed clock. Now you say, "Well we found out before that the thing to do is to coast along a straight line at a uniform speed chosen so that we arrive at B exactly 100 seconds later. If we don't go along a straight line it takes more speed, and our watch is slowed down." But wait! That was before we took gravity into account. Isn't it better to curve upward a little bit and then come down? Then during part of the time we are higher up and our watch will run a little faster? It is, indeed. If you solve the mathematical problem of adjusting the curve of the motion so that the elapsed time of the moving watch is the most it can possibly be, you will find that the motion is a parabola—the same curve followed by something that moves on a free ballistic path in the gravitational field, as in Figure 6-19. Therefore the law of motion in a gravitational field can also be stated: *An object always moves from one place to another so that a clock carried on it gives a longer time than it would on any other possible trajectory*—with, of course, the same starting and finishing conditions. The time measured by a moving clock is often called its "proper time." In free fall, the trajectory makes the proper time of an object a maximum.

Let's see how this all works out. We begin with Eq. (6.5) which says that the *excess* rate of the moving watch is

$$\frac{\omega_0 g H}{c^2}. \tag{6.13}$$

Besides this, we have to remember that there is a correction of the opposite sign for the speed. For this effect we know that

$$\omega = \omega_0 \sqrt{1 - v^2/c^2}.$$

Although the principle is valid for any speed, we take an example in which the speeds are always much less than c. Then we can write this equation as

$$\omega = \omega_0(1 - v^2/2c^2),$$

and the defect in the rate of our clock is

$$- \omega_0 \frac{v^2}{2c^2}. \tag{6.14}$$

Curved Space

Combining the two terms in (6.13) and (6.14) we have that

$$\Delta\omega = \frac{\omega_0}{c^2}\left(gH - \frac{v^2}{2}\right).$$ (6.15)

Such a frequency shift of our moving clock means that if we measure a time dt on a fixed clock, the moving clock will register the time

$$dt\left[1 + \left(\frac{gH}{c^2} - \frac{v^2}{2c^2}\right)\right].$$ (6.16)

The total time excess over the trajectory is the integral of the extra term with respect to time, namely

$$\frac{1}{c^2}\int\left(gH - \frac{v^2}{2}\right)\,dt,$$ (6.17)

which is supposed to be a maximum.

The term gH is just the gravitational potential ϕ. Suppose we multiply the whole thing by a constant factor $-mc^2$, where m is the mass of the object. The constant won't change the condition for the maximum, but the minus sign will just change the maximum to a minimum. Equation (6.16) then says that the object will move so that

$$\int\left(\frac{mv^2}{2} - m\phi\right)\,dt = \text{a minimum.}$$ (6.18)

But now the integrand is just the difference of the kinetic and potential energies. And if you look in Chapter 19 of Volume II* you will see that when we discussed the principle of least action we showed that Newton's laws for an object in any potential could be written exactly in the form of Eq. (6.18).

6-9 Einstein's theory of gravitation

Einstein's form of the equations of motion—that the proper time should be a maximum in curved space-time—gives the same results

---

* of the original *Lectures on Physics*.

as Newton's laws for low velocities. As he was circling around the earth, Gordon Cooper's watch was reading later than it would have in any other path you could have imagined for his satellite.*

So the law of gravitation can be stated in terms of the ideas of the geometry of space-time in this remarkable way. The particles always take the longest proper time—in space-time a quantity analogous to the "shortest distance." That's the law of motion in a gravitational field. The great advantage of putting it this way is that the law doesn't depend on any coordinates, or any other way of defining the situation.

Now let's summarize what we have done. We have given you two laws for gravity:

(1) How the geometry of space-time changes when matter is present—namely, that the curvature expressed in terms of the excess radius is proportional to the mass inside a sphere, Eq. (6.3).

(2) How objects move if there are only gravitational forces—namely, that objects move so that their proper time between two end conditions is a maximum.

Those two laws correspond to similar pairs of laws we have seen earlier. We originally described motion in a gravitational field in terms of Newton's inverse square law of gravitation and his laws of motion. Now laws (1) and (2) take their places. Our new pair of laws also correspond to what we have seen in electrodynamics. There we had our law—the set of Maxwell's equations—which determines the fields produced by charges. It tells how the character of "space" is changed by the presence of charged matter, which is what law (1) does for gravity. In addition, we had a law about how

---

* Strictly speaking it is only a *local* maximum. We should have said that the proper time is larger than for any *nearby* path. For example, the proper time on an elliptical orbit around the earth need not be longer than on a ballistic path of an object which is shot to a great height and falls back down.

Curved Space

particles move in the given fields—$d(m\mathbf{v})/dt = q(\mathbf{E} + \mathbf{v} \times \mathbf{B})$. This, for gravity, is done by law (2).

In the laws (1) and (2) you have a precise statement of Einstein's theory of gravitation—although you will usually find it stated in a more complicated mathematical form. We should, however, make one further addition. Just as time scales change from place to place in a gravitational field, so do the length scales. Rulers change lengths as you move around. It is impossible with space and time so intimately mixed to have something happen with time that isn't in some way reflected in space. Take even the simplest example: You are riding past the earth. What is "*time*" from *your* point of view is partly space from *our* point of view. So there must also be changes in space. It is the entire *space-time* which is distorted by the presence of matter, and this is more complicated than a change only in time scale. However, the rule that we gave in Eq. (6-3) is enough to determine completely all the laws of gravitation, provided that it is understood that this rule about the curvature of space applies not only from one man's point of view but is true for everybody. Somebody riding by a mass of material sees a different mass content because of the kinetic energy he calculates for its motion past him, and he must include the mass corresponding to that energy. The theory must be arranged so that everybody—no matter how he moves—will, when he draws a sphere, find that the excess radius is $G/3c^2$ times the total mass (or, better, $G/3c^4$ times the total energy content) inside the sphere. That this law—law (1)—should be true in any moving system is one of the great laws of gravitation, called *Einstein's field equation*. The other great law is (2)—that things must move so that the proper time is a maximum—and is called *Einstein's equation of motion*.

To write these laws in a complete algebraic form, to compare them with Newton's laws, or to relate them to electrodynamics is difficult mathematically. But it is the way our most complete laws of the physics of gravity look today.

Although they gave a result in agreement with Newton's mechanics for the simple example we considered, they do not always do so. The three discrepancies first derived by Einstein have been experimentally confirmed: The orbit of Mercury is not a fixed ellipse; starlight passing near the sun is deflected twice as much as you would think; and the rates of clocks depend on their location in a gravitational field. Whenever the predictions of Einstein have been found to differ from the ideas of Newtonian mechanics, Nature has chosen Einstein's.

Let's summarize everything that we have said in the following way. First, time and distance rates depend on the place in space you measure them and on the time. This is equivalent to the statement that space-time is curved. From the measured area of a sphere we can define a predicted radius, $\sqrt{A/4\pi}$, but the actual measured radius will have an excess over this which is proportional (the constant is G/c^2) to the total mass contained inside the sphere. This fixes the exact degree of the curvature of space-time. And the curvature must be the same no matter who is looking at the matter or how it is moving. Second, particles move on "straight lines" (trajectories of maximum proper time) in this curved space-time. This is the content of Einstein's formulation of the laws of gravitation.

Index

About Richard Feynman

Born in 1918 in Brooklyn, Richard P. Feynman received his Ph.D. from Princeton in 1942. Despite his youth, he played an important part in the Manhattan Project at Los Alamos during World War II. Subsequently, he taught at Cornell and at the California Institute of Technology. In 1965 he received the Nobel Prize in Physics, along with Sin-Itero Tomanaga and Julian Schwinger, for his work in quantum electrodynamics.

Dr. Feynman won his Nobel Prize for successfully resolving problems with the theory of quantum electrodynamics. He also created a mathematical theory that accounts for the phenomenon of superfluidity in liquid helium. Thereafter, with Murray Gell-Mann, he did fundamental work in the area of weak interactions such as beta decay. In later years Feynman played a key role in the development of quark theory by putting forward his parton model of high energy proton collision processes.

Beyond these achievements, Dr. Feynman introduced basic new computational techniques and notations into physics, above all, the ubiquitous Feynman diagrams that, perhaps more than any other formalism in recent scientific history, have changed the way in which basic physical processes are conceptualized and calculated.

Feynman was a remarkably effective educator. Of all his numerous awards, he was especially proud of the Oersted Medal for Teaching which he won in 1972. *The Feynman Lectures on Physics*, originally published in 1963, were described by a reviewer in *Scientific American* as "tough, but nourishing and full of flavor. After 25 years it is *the* guide for teachers and for the best of beginning students." In order to increase the understanding of physics among the lay public, Dr. Feynman wrote *The Character of Physical Law* and *Q.E.D.: The Strange Theory of Light and Matter*. He also authored a number of advanced publications that have become classic references and textbooks for researchers and students.

Richard Feynman was a constructive public man. His work on the Challenger commission is well known, especially his famous demonstration of the susceptibility of the O-rings to cold, an elegant experiment which required nothing more than a glass of ice water. Less well known were Dr. Feynman's efforts on the California State Curriculum Committee in the 1960s where he protested the mediocrity of textbooks.

A recital of Richard Feynman's myriad scientific and educational accomplishments cannot adequately capture the essence of the man. As any reader of even his most technical publications knows, Feynman's lively and multi-sided personality shines through all his work. Besides being a physicist, he was at various times a repairer of radios, a picker of locks, an artist, a dancer, a bongo player, and even a decipherer of Mayan hieroglyphics. Perpetually curious about his world, he was an exemplary empiricist.

Richard Feynman died on February 15, 1988, in Los Angeles.